SYSTEM
SIMULATION

SECOND EDITION

SYSTEM
SIMULATION

GEOFFREY GORDON

IBM Corporation

PRENTICE-HALL, INC., ENGLEWOOD CLIFFS, NEW JERSEY 07632

Library of Congress Cataloging in Publication Data

GORDON, GEOFFREY.
    System simulation.

    Includes bibliographies and index.
    1. Digital computer simulation.   2. System
analysis.   I. Title.
QA76.5.G63 1978   001.4'24   77-24579
ISBN   0-13-881797-9

© 1978, 1969 by Prentice-Hall, Inc., Englewood Cliffs, New Jersey   07632

10   9   8   7

Printed in the United States of America

PRENTICE-HALL INTERNATIONAL, INC., *London*
PRENTICE-HALL OF AUSTRALIA PTY. LIMITED, *Sydney*
PRENTICE-HALL OF CANADA, LTD., *Toronto*
PRENTICE-HALL OF INDIA PRIVATE LIMITED, *New Delhi*
PRENTICE-HALL OF JAPAN, INC., *Tokyo*
PRENTICE-HALL OF SOUTHEAST ASIA PTE. LTD., *Singapore*
WHITEHALL BOOKS LIMITED, *Wellington, New Zealand*

CONTENTS

2

SYSTEM STUDIES *21*

3

SYSTEM SIMULATION *38*

4

CONTINUOUS SYSTEM SIMULATION *58*

5

SYSTEM DYNAMICS *82*

6

PROBABILITY CONCEPTS IN SIMULATION *112*

7

ARRIVAL PATTERNS AND SERVICE TIMES *144*

8

DISCRETE SYSTEM SIMULATION *173*

INTRODUCTION TO GPSS *197*

10

GPSS EXAMPLES *222*

11

INTRODUCTION TO SIMSCRIPT *246*

PREFACE

The application of simulation continues to expand, both in terms of the extent to which simulation is used and the range of applications. This second edition of *System Simulation* reflects the changes.

The ties between analog and digital simulation, represented by digital-analog simulators, have become a matter of history. The section originally devoted to a specific digital-analog language has, therefore, been omitted. On the other hand, interest in socio-economic problems has greatly expanded the type of study originally described as Industrial Dynamics. The chapter on System Dynamics is devoted to this broader topic. The DYNAMO programming language was a pioneer in this type of study. It is still an active language, but other continuous-system simulation languages are now extensively used for the same purpose. The chapter devoted specifically to DYNAMO has, therefore, been omitted. Similarly, while FORTRAN is still being used for simulation programming, the general availability of programming languages designed for simulation makes the chapter that was devoted to that language unnecessary.

The development of discrete-system simulation languages continues. In fact, there are now languages that bridge the gap between continuous and discrete system simulation. GPSS and SIMSCRIPT, however, remain prominent. They have been retained as representatives of the dominant techniques for implementing discrete-system simulation.

The prime purpose of the book remains to provide material for a one-semester course introducing students to the topic and techniques of system simulation.

Some familiarity with programming concepts is assumed, but no knowledge of any particular programming language is needed as a prerequisite. In addition, familiarity with the concepts of statistics and probability theory will be helpful, although two chapters review the required theory. Many areas of application are discussed, but the technical details are kept simple so that the book will be useful to students of different backgrounds. A course based on the text could be introduced at the senior undergraduate or graduate level in many disciplines.

The first chapter discusses the types of models that are basic to any simulation. The second chapter considers the topic of organizing a system study. The nature of the simulation technique is then introduced in Chapter 3. Continuous-system simulation is examined in Chapter 4, and its application to System Dynamics studies is illustrated in Chapter 5. Chapters 6 and 7 are the ones that introduce the necessary statistics and probability theory.

The application of discrete-system simulation is demonstrated in Chapter 8, using a hand-worked example of a simple telephone-system. Two chapters are then devoted to GPSS, followed by two chapters describing SIMSCRIPT. In both cases, the first of the two chapters is self-contained, and would be sufficient for students needing only a brief introduction. The second chapter in each case includes, as a worked-out example, the same telephone-system problem presented in Chapter 8, thus giving an opportunity to compare the three approaches.

Chapter 13 discusses the programming techniques involved in the design and construction of simulation programming systems. Its prime interest will be to students concerned with the evaluation and design of simulation languages. The last chapter, also, is of a more technical nature, since it discusses the topic of analyzing statistical outputs of simulation runs.

I would like to thank CACI, Inc., for providing the SIMSCRIPT listings used in Chapters 11 and 12, and Dr. E. C. Russell for his generous help in both correcting the SIMSCRIPT examples and reviewing those chapters.

GEOFFREY GORDON

1

SYSTEM MODELS

1-1

The Concepts of a System

The term *system* is used in such a wide variety of ways that it is difficult to produce a definition broad enough to cover the many uses and, at the same time, concise enough to serve a useful purpose, (6), (12), and (20).[1] We begin, therefore, with a simple definition of a system and expand upon it by introducing some of the terms that are commonly used when discussing systems. A *system* is defined as an aggregation or assemblage of objects joined in some regular interaction or inter-dependence. While this definition is broad enough to include static systems, the principal interest will be in dynamic systems where the interactions cause changes over time.

As an example of a conceptually simple system, consider an aircraft flying under the control of an autopilot (see Fig. 1-1). A gyroscope in the autopilot detects the difference between the actual heading and the desired heading. It sends a signal to move the control surfaces. In response to the control surface movement, the airframe steers toward the desired heading.

As a second example, consider a factory that makes and assembles parts into a product (see Fig. 1-2). Two major components of the system are the fabrication

[1]Parenthetical numbers in text refer to items in bibliography at end of chapter.

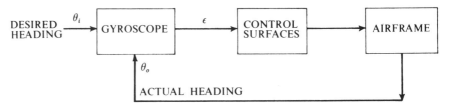

Figure 1-1. An aircraft under autopilot control.

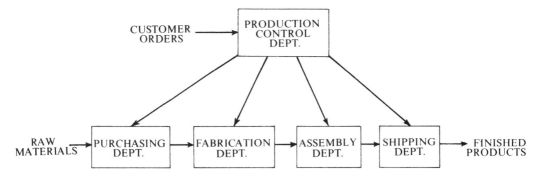

Figure 1-2. A factory system.

department making the parts and the assembly department producing the products. A purchasing department maintains a supply of raw materials and a shipping department dispatches the finished products. A production control department receives orders and assigns work to the other departments.

In looking at these systems, we see that there are certain distinct objects, each of which possesses properties of interest. There are also certain interactions occurring in the system that cause changes in the system. The term *entity* will be used to denote an object of interest in a system; the term *attribute* will denote a property of an entity. There can, of course, be many attributes to a given entity. Any process that causes changes in the system will be called an *activity*. The term *state of the system* will be used to mean a description of all the entities, attributes, and activities as they exist at one point in time. The progress of the system is studied by following the changes in the state of the system.

In the description of the aircraft system, the entities of the system are the airframe, the control surfaces, and the gyroscope. Their attributes are such factors as speed, control surface angle, and gyroscope setting. The activities are the driving of the control surfaces and the response of the airframe to the control surface movements. In the factory system, the entities are the departments, orders, parts, and products. The activities are the manufacturing processes of the departments.

Attributes are such factors as the quantities for each order, type of part, or number of machines in a department.

Figure 1-3 lists examples of what might be considered entities, attributes, and activities for a number of other systems. If we consider the movement of cars as a traffic system, the individual cars are regarded as entities, each having as attributes its speed and distance traveled. Among the activities is the driving of a car. In the case of a bank system, the customers of the bank are entities with the balances of their accounts and their credit statuses as attributes. A typical activity would be the action of making a deposit. Other examples are shown in Fig. 1-3.

SYSTEM	ENTITIES	ATTRIBUTES	ACTIVITIES
TRAFFIC	CARS	SPEED DISTANCE	DRIVING
BANK	CUSTOMERS	BALANCE CREDIT STATUS	DEPOSITING
COMMUNICATIONS	MESSAGES	LENGTH PRIORITY	TRANSMITTING
SUPERMARKET	CUSTOMERS	SHOPPING LIST	CHECKING-OUT

Figure 1-3. Examples of systems.

The figure does not show a complete list of all entities, attributes, and activities for the systems. In fact, a complete list cannot be made without knowing the purpose of the system description. Depending upon that purpose, various aspects of the system will be of interest and will determine what needs to be identified.

1-2

System Environment

A system is often affected by changes occurring outside the system. Some system activities may also produce changes that do not react on the system. Such changes occurring outside the system are said to occur in the *system environment*. An important step in modeling systems is to decide upon the boundary between the system and its environment. The decision may depend upon the purpose of the study.

In the case of the factory system, for example, the factors controlling the arrival of orders may be considered to be outside the influence of the factory and therefore part of the environment. However, if the effect of supply on demand is to be considered, there will be a relationship between factory output and arrival of orders, and this relationship must be considered an activity of the system. Similarly, in the case of a bank system, there may be a limit on the maximum interest rate that can be paid. For the study of a single bank, this would be regarded as a constraint imposed by the environment. In a study of the effects of monetary laws on the banking industry, however, the setting of the limit would be an activity of the system.

The term *endogenous* is used to describe activities occurring within the system and the term *exogenous* is used to describe activities in the environment that affect the system. A system for which there is no exogenous activity is said to be a *closed* system in contrast to an *open* system which does have exogenous activities.

1-3
Stochastic Activities

One other distinction that needs to be drawn between activities depends upon the manner in which they can be described. Where the outcome of an activity can be described completely in terms of its input, the activity is said to be *deterministic*. Where the effects of the activity vary randomly over various possible outcomes, the activity is said to be *stochastic*.

The randomness of a stochastic activity would seem to imply that the activity is part of the system environment since the exact outcome at any time is not known. However, the random output can often be measured and described in the form of a probability distribution. If, however, the *occurrence* of the activity is random, it will constitute part of the environment. For example, in the case of the factory, the time taken for a machining operation may need to be described by a probability distribution but machining would be considered to be an endogenous activity. On the other hand, there may be power failures at random intervals of time. They would be the result of an exogenous activity.

If an activity is truly stochastic, there is no known explanation for its randomness. Sometimes, however, when it requires too much detail or is just too much trouble to describe an activity fully, the activity is represented as stochastic. For example, in modeling elevator service in a building, the re-entry of people into the elevator, after they have been taken to a floor, could be connected with their having left the elevator, by assigning the time they stay on the floor. In most models, however, leaving and re-entry would be treated as separate stochastic activities, connected only by the fact that the mean rates at which they transfer people are equal.

Assembling the data for a model will often involve an element of uncertainty that arises from sampling or experimental error. A value for some attribute of a model, which is known to be fixed, must be selected from a number of recorded values that contain random errors. Deciding on the best estimate is a statistical exercise. Usually, an arithmetic average will be considered sufficiently accurate.

1-4

Continuous and Discrete Systems

The aircraft and factory systems used as examples in Sec. 1-1 respond to environmental changes in different ways. The movement of the aircraft occurs smoothly, whereas the changes in the factory occur discontinuously. The ordering of raw materials or the completion of a product, for example, occurs at specific points in time.

Systems such as the aircraft, in which the changes are predominantly smooth, are called *continuous systems*. Systems like the factory, in which changes are predominantly discontinuous, will be called *discrete systems*. Few systems are wholly continuous or discrete. The aircraft, for example, may make discrete adjustments to its trim as altitude changes, while, in the factory example, machining proceeds continuously, even though the start and finish of a job are discrete changes. However, in most systems one type of change predominates, so that systems can usually be classified as being continuous or discrete.

The complete aircraft system might even be regarded as a discrete system. If the purpose of studying the aircraft were to follow its progress along its scheduled route, with a view, perhaps, to studying air traffic problems, there would be no point in following precisely *how* the aircraft turns. It would be sufficiently accurate to treat changes of heading at scheduled turning points as being made instantaneously, and so regard the system as being discrete.

In addition, in the factory system, if the number of parts is sufficiently large, there may be no point in treating the number as a discrete variable. Instead, the number of parts might be represented by a continuous variable with the machining activity controlling the rate at which parts flow from one state to another. This is, in fact, the approach of a modeling technique called System Dynamics, which will be discussed in Chap. 5.

There are also systems that are intrinsically continuous but information about them is only available at discrete points in time. These are called *sampled-data* systems, (15). The study of such systems includes the problem of determining the effects of the discrete sampling, especially when the intention is to control the system on the basis of information gathered by the sampling.

This ambiguity in how a system might be represented illustrates an important point. The description of a system, rather than the nature of the system itself, determines what type of model will be used. A distinction needs to be made because, as will be discussed later, the general programming methods used to simulate continuous and discrete models differ. However, no specific rules can be given as to how a particular system is to be represented. The purpose of the model, coupled with the general principle that a model should not be more complicated than is needed, will determine the level of detail and the accuracy with which a model needs to be developed. Weighing these factors and drawing on the experience of knowledgeable people will decide the type of model that is needed.

1-5
System Modeling

To study a system, it is sometimes possible to experiment with the system itself. The objective of many system studies, however, is to predict how a system will perform before it is built. Clearly, it is not feasible to experiment with a system while it is in this hypothetical form. An alternative that is sometimes used is to construct a number of prototypes and test them, but this can be very expensive and time-consuming. Even with an existing system, it is likely to be impossible or impractical to experiment with the actual system. For example, it is not feasible to study economic systems by arbitrarily changing the supply and demand of goods. Consequently, system studies are generally conducted with a model of the system. For the purpose of most studies, it is not necessary to consider all the details of a system; so a model is not only a substitute for a system, it is also a simplification of the system, (14).

We define a *model* as the body of information about a system gathered for the purpose of studying the system.[2] Since the purpose of the study will determine the nature of the information that is gathered, there is no unique model of a system. Different models of the same system will be produced by different analysts interested in different aspects of the system or by the same analyst as his understanding of the system changes.

The task of deriving a model of a system may be divided broadly into two subtasks: establishing the model structure and supplying the data. Establishing the structure determines the system boundary and identifies the entities, attributes, and

---

[2]In the case of a physical model, the information is embodied in the properties of the model, in contrast to the symbolic representation in a mathematical model.

activities of the system. The data provide the values the attributes can have and define the relationships involved in the activities. The two jobs of creating a structure and providing the data are defined as parts of one task rather than as two separate tasks, because they are usually so intimately related that neither can be done without the other. Assumptions about the system direct the gathering of data, and analysis of the data confirms or refutes the assumptions. Quite often, the data gathered will disclose an unsuspected relationship that changes the model structure.

To illustrate this process, consider the following description of a supermarket, (3).

Shoppers needing *several items* of shopping *arrive* at a supermarket. They *get a basket*, if one is *available*, carry out their *shopping*, and then *queue* to *check-out* at one of the *several counters.*

After checking-out, they *return* the *basket* and *leave.*

Certain words have been italicized because they are considered to be key words that point out some feature of the system that must be reflected in the model. Essentially the same description is rewritten in Fig. 1-4 to identify the entities,

ENTITY	ATTRIBUTE	ACTIVITY
SHOPPER	NO. OF ITEMS	
		ARRIVE
		GET
BASKET	AVAILABILITY	
		SHOP
		QUEUE
		CHECK-OUT
COUNTER	NUMBER OCCUPANCY	
		RETURN
		LEAVE

Figure 1-4. Elements of a supermarket model.

attributes, and activities. Notice that the concept of a supermarket as a whole does not appear as an entity. It defines the system boundary and therefore distinguishes between the system and its environment. The arrival of customers in this description of the system will be regarded as an exogenous activity affecting the system from the environment. If, in contrast, the study objectives include analyzing the effects of car parking facilities on supermarket business, the boundary of the system would need to include the parking lot. The arrival of a customer in the supermarket depends upon finding a parking space, which can depend upon the departure of customers. Customer arrivals in the supermarket then become an endogenous activity; the arrival of cars becomes an exogenous activity.

Other decisions about the system study objectives are implied in the model. The number of items of shopping is represented as an attribute of the shopper, but no distinction has been made about the type of item. Secondly, no provision has been made in the system model for the effects of congestion on shopping time. If these decisions are not in keeping with the study objectives, another form of model must be used. In the first case, where type of item is to be distinguished, it is necessary to define several attributes for each customer, one for each type of item to be purchased. In the second case, where allowance for congestion must be made, two approaches could be taken. It may be necessary to introduce new entities representing the various sections of the supermarket and establish as attributes the number of customers they can serve simultaneously. Alternatively, the activity of shopping could be represented by a function in which shopping time depends upon the number of shoppers in the supermarket.

It is not suggested that Fig. 1-4 represents a formal process by which a transition can be made from a verbal description of a system to the structure of a model. It merely illustrates the process involved in forming a model.

1-6
Types of Models

Models used in system studies have been classified in many ways, (8), (9), (16), and (17). The classifications that will be used here are illustrated in Fig. 1-5. Models will first be separated into *physical models* or *mathematical models*.

Physical models are based on some analogy between such systems as mechanical and electrical, (4) and (11), or electrical and hydraulic, (10). In a physical model of a system, the system attributes are represented by such measurements as a voltage or the position of a shaft. The system activities are reflected in the physical laws that drive the model. For example, the rate at which the shaft of a direct current motor turns depends upon the voltage applied to the motor. If the applied voltage is used to represent the velocity of a vehicle, then the number of revolutions of the shaft is a measure of the distance the vehicle has traveled; the higher the voltage, or velocity, the greater is the buildup of revolutions, or distance covered, in a given time.

Mathematical models, of course, use symbolic notation and mathematical equations to represent a system. The system attributes are represented by variables,

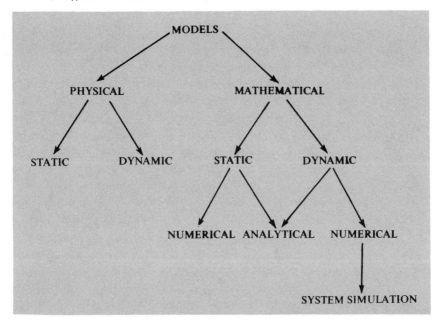

Figure 1-5. Types of models.

and the activities are represented by mathematical functions that interrelate the variables.

A second distinction will be between *static models* and *dynamic models*. Static models can only show the values that system attributes take when the system is in balance. Dynamic models, on the other hand, follow the changes over time that result from the system activities.

In the case of mathematical models, a third distinction is the technique by which the model is "solved," that is, actual values are assigned to system attributes. A distinction is made between *analytical* and *numerical* methods. Applying analytical techniques means using the deductive reasoning of mathematical theory to solve a model. In practice, only certain forms of equations can be solved. Using analytical techniques, therefore, is a matter of finding the model that can be solved and best fits the system being studied. For example, linear differential equations can be solved. Knowing this, an engineer who restricts the description of a system to that form will derive a model that can be solved analytically.

Numerical methods involve applying computational procedures to solve equations. To be strictly accurate, any assignment of numerical values that uses mathematical tables involves numerical methods, since tables are derived numerically. The distinction being drawn here is that analytical methods produce solutions in tractable form, meaning a form where values can be assigned from available tables. Making use of an analytical solution may, in fact, require a considerable amount of computation. For example, the solution may be derived in the form of a complicated integral which then needs to be expanded as a power series for evaluation. However, mathematical theory for making such expansions exists, and, in principle, any degree of accuracy in the solution is obtainable if sufficient effort is expended.

As will be discussed more fully in Chap. 3, system simulation is considered to be a numerical computation technique used in conjunction with dynamic mathematical models. Simulation models, therefore, are shown under that heading in Fig. 1-5.

Yet another distinction by which models are often classified is between deterministic and stochastic: the latter term meaning that there are random processes in the system. As will be discussed in Chap. 14, the introduction of stochastic processes in a simulation model complicates the task of interpreting results, and it increases the amount of work to be done. It does not, however, change the basic technique by which simulation is applied, so this distinction has not been made in Fig. 1-5.

1-7
Static Physical Models

The best known examples of physical models are scale models. In shipbuilding, making a scale model provides a simple way of determining the exact measurements of the plates covering the hull, rather than having to produce drawings of complicated, three-dimensional shapes. Scientists have used models in which spheres represent atoms, and rods or specially shaped sheets of metal connect the spheres to represent atomic bonds. A model of this nature played an important role in the deciphering of the DNA molecule, work that was the subject of a Nobel Prize award, (19). These models are static physical models. They are sometimes said to be *iconic* models, a term meaning "look-alike" (7).

Scale models are also used in wind tunnels and water tanks (1) in the course of designing aircraft and ships. Although air is blown over the model, or the model is pulled through the water, these are static physical models because the measurements that are taken represent attributes of the system being studied under one set of equilibrium conditions. In this case, the measurements do not translate directly into system attribute values. Well known laws of similitude are used to convert measurements on the scale model to the values that would occur in the real system, (5) and (13).

Sometimes, a static physical model is used as a means of solving equations with particular boundary conditions. There are many examples in the field of mathematical physics where the same equations apply to different physical phenomena. For example, the flow of heat and the distribution of electric charge through space can be related by common equations. In general, these equations can only be solved for simple-shaped bodies. In practice, solutions are needed for specific, complicated shapes. The distribution of heat in a body can be predicted by enclosing a space that has the same shape as the body, and measuring the charge in the space when the surface of the space has been electrified in a manner that reflects the way heat will be injected into the body, (2).

1-8

Dynamic Physical Models

Dynamic physical models rely upon an analogy between the system being studied and some other system of a different nature, the analogy usually depending upon an underlying similarity in the forces governing the behavior of the systems. To illustrate this type of physical model, consider the two systems shown in Fig. 1-6. Figure 1-6(a) represents a mass that is subject to an applied force $F(t)$ varying

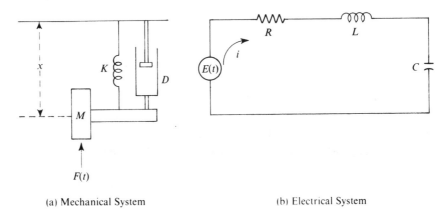

(a) Mechanical System   (b) Electrical System

Figure 1-6. Analogy between mechanical and electrical systems.

with time, a spring whose force is proportional to its extension or contraction, and a shock absorber that exerts a damping force proportional to the velocity of the mass. The system might, for example, represent the suspension of an automobile wheel when the automobile body is assumed to be immobile in a vertical direction. It can be shown that the motion of the system is described by the following differential equation:[3]

$$M\ddot{x} + D\dot{x} + Kx = KF(t)$$

[3]The derivation of this equation will be given in Sec. 4-2.

where x is the distance moved,

 M is the mass,

 K is the stiffness of the spring,[4]

 D is the damping factor of the shock absorber.

Figure 1-6(b) represents an electrical circuit with an inductance L, a resistance R, and a capacitance C, connected in series with a voltage source that varies in time according to the function $E(t)$. If q is the charge on the capacitance, it can be shown that the behavior of the circuit is governed by the following differential equation:

$$L\ddot{q} + R\dot{q} + \frac{q}{C} = \frac{E(t)}{C}$$

Inspection of these two equations shows that they have exactly the same form and that the following equivalences occur between the quantities in the two systems:

Displacement	x	Charge	q
Velocity	$\dot{x}$	Current	$I\,(=\dot{q})$
Force	F	Voltage	E
Mass	M	Inductance	L
Damping factor	D	Resistance	R
Spring stiffness	K	1/Capacitance	$1/C$

The mechanical system and the electrical system are analogs of each other, and the performance of either can be studied with the other. In practice, it is simpler to modify the electrical system than to change the mechanical system, so it is more likely that the electrical system will have been built to study the mechanical system. If, for example, a car wheel is considered to bounce too much with a particular suspension system, the electrical model will demonstrate this fact by showing that the charge (and, therefore, the voltage) on the condenser oscillates excessively. To predict what effect a change in the shock absorber or spring will have on the performance of the car, it is only necessary to change the values of the resistance or

---

[4]The constant K is included on the right-hand side of the equation as a matter of convenience. To demonstrate the characteristic behavior of a system, it is customary to show its response to a step-function of force, that is, a steady force applied when the body is stationary. The step-function is usually taken to be of unit magnitude.

If the system is stable, its response settles to a steady value. With K on the right-hand side of the equation, the steady value is always 1. This makes it possible to draw the responses for different values of coefficients on a common graph, as will be done in Fig. 1-9. Interpreting $F(t)$ as being a unit step-function means that the applied force is actually a step-function of magnitude K.

condenser in the electrical circuit and observe the effect on the way the voltage varies.

If, in fact, the mechanical system were as simple as illustrated, it could be studied by solving the mathematical equation derived in establishing the analogy. However, effects can easily be introduced that would make the mathematical equation difficult to solve. For example, if the motion of the wheel is limited by physical stops, a non-linear equation that is difficult to solve will be needed to describe the system. It is easy to model the effect electrically by placing limits on the voltage that can exist on the capacitance.

1-9

Static Mathematical Models

A static model gives the relationships between the system attributes when the system is in equilibrium. If the point of equilibrium is changed by altering any of the attribute values, the model enables the new values for all the attributes to be derived but does not show the way in which they changed to their new values.

For example, in marketing a commodity there is a balance between the supply and demand for the commodity. Both factors depend upon price: a simple *market model* will show what is the price at which the balance occurs.

Demand for the commodity will be low when the price is high, and it will increase as the price drops. The relationship between demand, denoted by Q, and price, denoted by P, might be represented by the straight line marked "Demand" in in Fig. 1-7.[5] On the other hand, the supply can be expected to increase as the price increases, because the suppliers see an opportunity for more revenue. Suppose supply, denoted by S, is plotted against price, and the relationship is the straight line marked "Supply" in Fig. 1-7. If conditions remain stable, the price will settle to the point at which the two lines cross, because that is where the supply equals the demand.

Since the relationships have been assumed linear, the complete market model can be written mathematically as follows:

$$Q = a - bP$$
$$S = c + dP$$
$$S = Q$$

---

[5]This description makes price the independent variable, and demand the dependent variable. However, Figs. 1-7 and 1-8 place price along the vertical axis in order to conform with the practice normally used in the literature of economics. See (18).

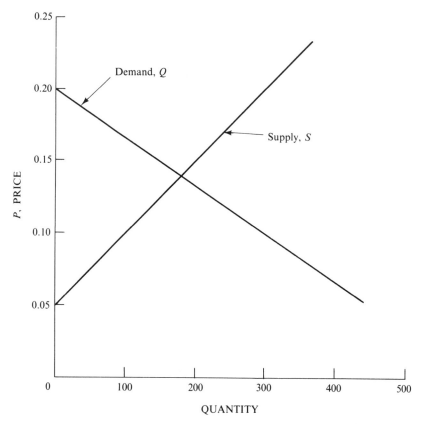

Figure 1-7. Linear market model.

The last equation states the condition for the market to be cleared; it says supply equals demand and, so, determines the price to which the market will settle.

For the model to correspond to normal market conditions in which demand goes down and supply increases as price goes up the coefficients b and d need to be positive numbers. For realistic, positive results, the coefficient a must also be positive. Figure 1-7 has been plotted for the following values of the coefficients:

$$a = \quad 600$$
$$b = 3{,}000$$
$$c = -100$$
$$d = 2{,}000$$

The fact that linear relationships have been assumed allows the model to be solved analytically. The equilibrium market price, in fact, is given by the following expression:

$$P = \frac{a - c}{b + d}$$

With the chosen values, the equilibrium price is 0.14, which corresponds to a supply of 180.

More usually, the demand will be represented by a *curve* that slopes downwards, and the supply by a curve that slopes upwards, as illustrated in Fig. 1-8. It may not

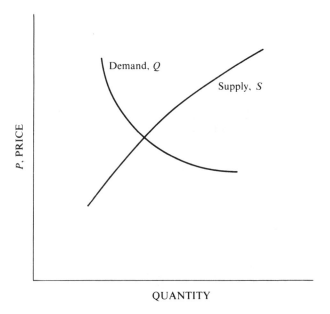

QUANTITY

Figure 1-8. Non-linear market model.

then be possible to express the relationships by equations that can be solved. Some numeric method is then needed to solve the equations. Drawing the curves to scale and determining graphically where they intersect is one such method.

In practice, it is difficult to get precise values for the coefficients of the model. Observations over an extended period of time, however, will establish the slopes (that is, the values of b and d) in the neighborhood of the equilibrium point, and, of course, actual experience will have established equilibrium prices under various conditions. The values depend upon economic factors, so the observations will usually attempt to correlate the values with the economy, allowing the model to be used as a means of forecasting changes in market conditions for anticipated economic changes.

1-10

Dynamic Mathematical Models

A dynamic mathematical model allows the changes of system attributes to be derived as a function of time. The derivation may be made with an analytical solution or with a numerical computation, depending upon the complexity of the model. The equation that was derived to describe the behavior of a car wheel is an example of a dynamic mathematical model; in this case, an equation that can be solved analytically. It is customary to write the equation in the form

$$\ddot{x} + 2\zeta\omega\dot{x} + \omega^2 x = \omega^2 F(t) \qquad (1\text{-}1)$$

where $2\zeta\omega = D/M$ and $\omega^2 = K/M$.

Expressed in this form, solutions can be given in terms of the variable ωt. Figure 1-9 shows how x varies in response to a steady force applied at time $t = 0$ as would occur, for instance, if a load were suddenly placed on the automobile. Solutions are shown for several values of ζ, and it can be seen that when ζ is less than 1, the motion is oscillatory.

The factor ζ is called the *damping ratio* and, when the motion is oscillatory, the

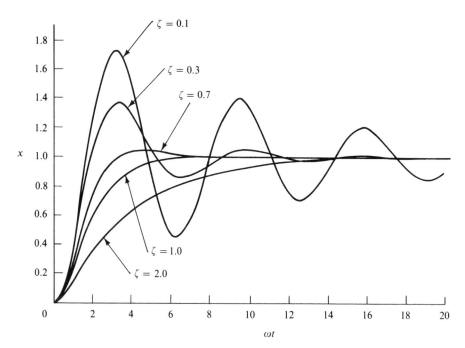

Figure 1-9. Solutions of second-order equations.

frequency of oscillation is determined from the formula

$$\omega = 2\pi f$$

where f is the number of cycles per second.

Suppose a case is selected as representing a satisfactory frequency and damping. The relationships given above between ζ, ω, M, K, and D show how to select the spring and shock absorber to get that type of motion. For example, the condition for the motion to occur without oscillation requires that $\zeta \geq 1$. It can be deduced from the definition of ζ and ω that the condition requires that $D^2 \geq 4MK$.

1-11

Principles Used in Modeling

It is not possible to provide rules by which mathematical models are built, but a number of guiding principles can be stated. They do not describe distinct steps carried out in building a model. They describe different viewpoints from which to judge the information to be included in the model.

(a) Block-building

The description of the system should be organized in a series of *blocks*. The aim in constructing the blocks is to simplify the specification of the interactions within the system. Each block describes a part of the system that depends upon a few, preferably one, input variables and results in a few output variables. The system as a whole can then be described in terms of the interconnections between the blocks. Correspondingly, the system can be represented graphically as a simple block diagram.

The description of a factory given in Fig. 1-2 is a typical example of a block diagram. Each department of the factory has been treated as a separate block, with the inputs and outputs being the work passed from department to department. The fact that the departments might occupy the same floor space and might use the same personnel or the same machines has been ignored.

(b) Relevance

The model should only include those aspects of the system that are relevant to the study objectives. As an example, if the factory system study aims to compare the effects of different operating rules on efficiency, it is not relevant to consider the hiring of employees as an activity. While irrelevant information in the model may not do any harm, it should be excluded because it increases the complexity of the model and causes more work in solving the model.

(c) Accuracy

The accuracy of the information gathered for the model should be considered. In the aircraft system, for example, the accuracy with which the movement of the aircraft is described depends upon the representation of the airframe. It may suffice to regard the airframe as a rigid body and derive a very simple relationship between control surface movement and aircraft heading, or it may be necessary to recognize the flexibility of the airframe and make allowance for vibrations in the structure. An engineer responsible for estimating the fuel consumption may be satisfied with the simple representation. Another engineer, responsible for considering the comfort of the passengers, needs to consider vibrations and will want the detailed description of the airframe.

(d) Aggregation

A further factor to be considered is the extent to which the number of individual entities can be grouped together into larger entities. The general manager of the factory may be satisfied with the description that has been given. The production control manager, however, will want to consider the shops of the departments as individual entities.

In some studies, it may be necessary to construct artificial entities through the process of aggregation. For example, an economic or social study will usually treat a population as a number of social classes and conduct a study as though each social class were a distinct entity.

Similar considerations of aggregation should be given to the representation of activities. For example, in studying a missile defense system, it may not be necessary to include the details of computing a missile trajectory for each firing. It may be sufficient to represent the outcome of many firings by a probability function.

Exercises

1-1   Extract from the following description the entities, attributes, and activities of the system. Ships arrive at a port. They dock at a berth if one is available; otherwise, they wait until one becomes available. They are unloaded by one of several work gangs whose size depends upon the ship's tonnage. A warehouse contains a new cargo for the ship. The ship is loaded and then departs. Suggest two exogenous events (other than arrivals) that may need to be taken into account.

1-2 Name three or four of the principal entities, attributes, and activities to be considered if you were to simulate the operation of (a) a gasoline filling station, (b) a cafeteria, (c) a barber shop.

1-3 A new bus route is to be added to a city, and the traffic manager is to determine how many extra buses will be needed. What are the three key attributes of the passengers and buses that he should consider? If the company manager wants to assess the effect of the new route on the transit system as a whole, how would you suggest he aggregate the features of the new line to form part of a total system model? Would you suggest a continuous or discrete model for the traffic manager and the general manager?

1-4 In the automobile wheel suspension system, it is found that the shock absorber damping force is not strictly proportional to the velocity of the wheel. There is an additional force component equal to D_2 times the acceleration of the wheel. Find the new conditions for ensuring that the wheel does not oscillate.

1-5 A woman does her shopping on Mondays, Wednesdays, and Fridays. If it is fine, she walks to the stores; otherwise, she takes a bus. She always takes the bus home. On Tuesdays, she visits her daughter, traveling there and back by bus. Assuming that information is available about the day of the week and the state of the weather, draw a flow chart of her movements.

1-6 In the aircraft system, suppose the control surface angle y is made to be A times the error signal. The response of the aircraft to the control surface is found to be $I\ddot{\theta}_o + D\dot{\theta}_o = Ky$. Find the conditions under which the aircraft motion is oscillatory.

1-7 Suppose the automobile body in the suspension system example is not stationary. Consider the body to have a mass of M_1, and assume that its motion is determined by the force of gravity and the reaction with the suspension system. Construct a model for the motions of the wheel and body.

Bibliography

1   ALLEN, J., *Scale Models in Hydraulic Engineering*, London: Longman Green & Co., 1947.

2   BARABASCHI, S., M. CONTI, L. GENTILINI, AND A. MATHIS, "A Heat Exchange Simulator," *Automatica*, II, no. 1 (1964), 1–13.

3   BLAKE, K., AND G. GORDON, "Systems Simulation with Digital Computers," *IBM Syst. J.*, III, no. 1 (1964), 14–20.

4   BLOCH, A., "Electromechanical Analogies and Their Use for the Analysis of

Mechanical and Electromechanical Systems," *J. Inst. Electr. Eng.*, XCII, no. 52 (1945), 157–169.

5 BRIDGEMAN, PERCY W., *Dimensional Analysis*, New York: AMS Press, Inc., 1976.

6 CHORAFAS, D. N., *Systems and Simulation*, New York: Academic Press, Inc., 1965.

7 CHURCHMAN, C. WEST, RUSSELL L. ACKOFF, AND E. LEONARD ARNOFF, *Introduction to Operations Research*, chap. 7, New York: John Wiley & Sons, Inc., 1957.

8 EMSHOFF, JAMES R., AND ROGER L. SISSON, *Design and Use of Computer Simulation Models*, New York: The Macmillan Company, 1970.

9 HIGHLAND, HAROLD JOSEPH, "A Taxonomy of Models," *Simuletter*, IV, no. 2 (1973), 10–17.

10 HUBBARD, P. G., "Applications of the Electrical Analogy in Fluid Mechanics Research," *Rev. Sci. Instrum.*, XX, no. 11 (1949), 802–807.

11 KEROPYAN, K. K., ed., *Electrical Analogues of Pin-Jointed Systems*, New York: The Macmillan Company, 1965.

12 KLIR, GEORGE J., *An Approach to General Systems Theory*, New York: Van Nostrand Reinhold Co., 1969.

13 LANGHAAR, HENRY LEWIS, *Dimensional Analysis and Theory of Models*, New York: John Wiley & Sons, Inc., 1951.

14 MIHRAM, DANIELLE, AND G. ARTHUR MIHRAM, "Human Knowledge, The Role of Models, Metaphors and Analogy," *International J. General Systems*, I, no. 1 (1974), 41–60.

15 RAGAZZINI, J. R., AND L. A. ZADEH, "The Analysis of Sample-Data Systems," *Trans. Am. Inst. Electr. Eng.*, LXXI, pt. II (1952), 225–232.

16 RIVETT, B. H., *Principles of Model Building: The Construction of Models for Decision Making*, New York: John Wiley & Sons, Inc., 1972.

17 RIVETT, PATRICK, AND RUSSELL LINCOLN ACKOFF, *Manager's Guide to Operations Research*, New York: John Wiley & Sons, Inc., 1963.

18 SAMUELSON, PAUL A., *Economics* (10th ed.), chap. 4, New York: McGraw-Hill Book Company, 1976.[6]

19 WATSON, JAMES D., *The Double Helix*, New York: Atheneum Publishers, 1968.

20 ZEIGLER, BERNARD P., *Theory of Modeling and Simulation*, New York: John Wiley & Sons, Inc., 1976.

[6]See footnote 5, this chapter.

2

SYSTEM STUDIES

2-1

Subsystems

We have given a simple definition of a system as a set of interacting objects. The objects might be considered the basic entities of the system, but, usually, the description of a system can be made at many levels of detail. It is customary to describe a system as consisting of interacting *subsystems*. Any subsystem might, itself, be considered a system consisting of subsystems at a still lower level of detail, and so on. A system study must begin by deciding on the level of subsystem detail to be used, (7).

The block-building principle, mentioned in Sec. 1-11, helps organize a system description by isolating subsystems and identifying their inputs and outputs. Each subsystem at a given level of detail is described as a block, giving relationships between the inputs and outputs. The relationships should be such that they are sufficient to determine the outputs from the inputs when the subsystem stands by itself; that is to say, there should be no need to use an endogenous variable within the block. The term "black box," borrowed from the engineering field, is often used to describe an element of this nature which gives an output in response to an input without any need, or ability, to know how the transformation is made.

Each subsystem has its own inputs and outputs and, standing by itself, its response can be derived from the relationship that defines the subsystem. The interactions that occur in a system arise from the fact that the outputs of some subsystems become the inputs of others: they become endogenous variables of the system.

In the same way that a system breaks down into subsystems, so also a model of a system breaks into *submodels*. When describing systems in terms of blocks, the terms *block*, *subsystem*, or *submodel* tend to be used interchangeably, as will occur in the subsequent discussion. In fact, when concentrating on a particular part of a system or its model, we will often drop the prefix "sub," and talk about a part of a system or model as being, itself, a system or model. The context should make it quite clear what is intended.

2-2
A Corporate Model

We will illustrate the interactions that occur in a system by looking at a model of a corporation. Models of this nature are called *corporate models*, and they are used by many corporations to help in various aspects of planning their operations, (8) and (13). Their main use is in helping to understand what conditions will prevail under different sets of assumptions.

Assume the corporation is in the manufacturing industry, and it is planning to produce and market some new product. A first level of detail might consider the

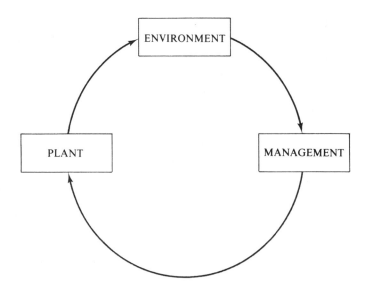

Figure 2-1. Major segments of a corporate model.

complete model as consisting of three parts, each of a different general nature. The corporation, as a whole, operates within an environment, which, as was defined in Sec. 1-2, includes all those factors influencing the corporation that are not under its direct control. Looking at the corporation itself, it falls into two major segments: the physical plant, which provides the means for production; and the management segment, representing the policy-making aspects of the corporation. These three major subsystems interact as shown in Fig. 2-1.

2-3

Environment Segment

Figure 2-2 shows how the environment subsystem might be further broken down. The major element is the market model, the set of relationships that determine what the demand for the product will be. Demand is the main output of this

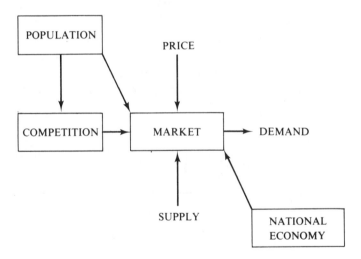

Figure 2-2. The environment segment.

model. Two major inputs are the price and the supply of the product. Several other factors can influence the market. The ones chosen here are the national economy, the population that purchases the product, and competition from other producers.

The national economy affects the market by influencing the amount of money available to consumers, (9). It was mentioned in Sec. 1-9, when discussing a simple market model, that changes in economic conditions would change the coefficients of that model. If the national economy is to be included in the present model, an output of this subsystem would be the coefficients to be used in the market model.

The obvious influence of population on the market is through knowing how many people are likely to buy the product, (10). An additional factor could be the distribution of the population, if the model is to go as far as considering where the product distribution outlets are to be placed. If a population subsystem is to be included, therefore, it will contain demographic information. It might also include, at a lower level of detail, an advertising subsystem, because the significant factor is really the number of people who decide to buy, and that is influenced by advertising.

Competition from other producers, or from similar products, is another factor to be considered, (1). If this is included, the competition model is likely to interact with the population model.

2-4

Production Segment

A breakdown of the production subsystem is shown in Fig. 2-3. The main input is the amount of labor and machinery assigned for production. The main output is the supply of the product. If it is decided to look at the market geographically, then a distribution subsystem will be needed to investigate how the location of warehouses should be matched to population distribution.

Production will be influenced by economic factors which might be divided into

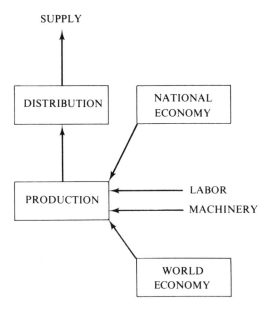

Figure 2-3. The production segment.

two categories. Wages affect production, through labor supply, and they are an important item in any national economy model. If a national economy model is to be included, therefore, it will interact with the production model. The world economy might also be important, if the cost of imported raw materials has to be considered, or if any of the production facilities are located overseas. Of course, if the product is sold overseas, the world economy model will also interact with the market model.

2-5

Management Segment

The management subsystem, further broken down, is shown in Fig. 2-4. The main inputs are the demand for the product, and the capital investment to be made in the business. The main outputs are the price to be set and the profit to be

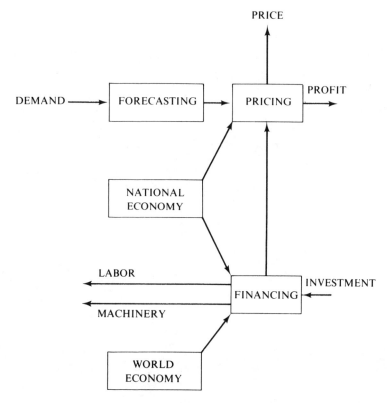

Figure 2-4. The management segment.

expected. A pricing model sets the price, and a financial model decides how the investment capital is to be divided between labor and machinery.

Another activity of the management is to predict the future demand for the product. A forecasting model may be needed to reflect the way the predictions are made, (14).

Both the national and the world economic conditions could influence the financial model through their effects on the money market.

2-6

The Full Corporate Model

Putting together the expansions of the three segments of Fig. 2-1 gives the full model of Fig. 2-5. We have now gone from three subsystems to ten, and, as has been intimated, it is possible to expand to a much larger number. Without knowing

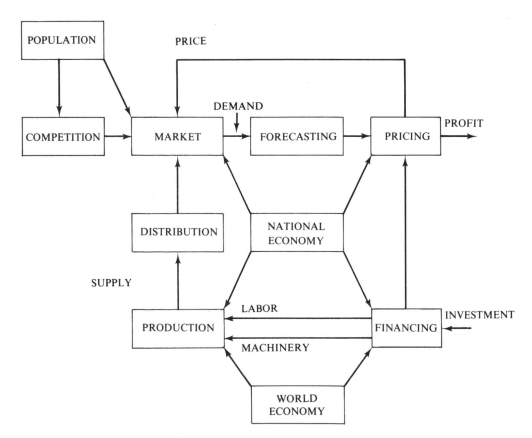

Figure 2-5. The full corporate model.

the purpose and objectives of the study, it is not possible to judge where the process of refinement should stop. We will assume we have gone far enough. Later in this chapter, we will analyze a simplified version of the model.

Our purposes here have been to demonstrate how interactions arise in a system, and show the process of refinement by which full models are constructed. As can be imagined, there are many judgments involved in deciding what shall be included in a model, judgments that usually need to be made by a person familiar with the system and the objectives of the study. Often, data analysis will help make the decision. Examining data and plotting one variable against another, or using more formal statistical analysis techniques such as correlation or regression analysis, may uncover a significant relationship, or show whether a proposed relationship is strong enough to be worth considering.

2-7
Types of System Study

Having developed a model, there are various ways we can use it to study a system. Generally, system studies are of three types: system analysis, system design, and what will be called system postulation.

Many studies, in fact, combine two or three of these aspects or alternate between them as the study proceeds. The term *system engineering* is frequently used to describe system studies where a combination of analysis and design is aimed at understanding, first, how an existing system works and then preparing system modifications to change the system behavior.

System analysis aims to understand how an existing system or a proposed system operates. The ideal situation would be that the investigator is able to experiment with the system itself. What is actually done is to construct a model of the system and investigate the behavior of the model. The results obtained are interpreted in terms of system performance.

In *system design* studies, the object is to produce a system that meets some specifications. Certain system parameters or components can be selected or planned by the designer, and, conceptually, he chooses a particular combination of components to construct a system. The proposed system is modeled and its performance predicted from knowledge of the model's behavior. If the predicted performance compares favorably with the desired performance, the design is accepted. Otherwise, the system is redesigned and the process repeated.

System postulation is characteristic of the way models are employed in social, economic, political, and medical studies, where the behavior of the system is known but the processes that produce the behavior are not. Hypotheses are made on a likely set of entities and activities that can explain the behavior. The study compares the response of the model based on these hypotheses with the known behavior. A reasonably good match naturally leads to the assumption that the structure of the model bears a resemblance to the actual system, and allows a system structure to be postulated. Very likely, the behavior of the model gives a better understanding of the system, possibly helping to formulate a refined set of hypotheses.

In the remainder of this chapter we will demonstrate these three general approaches to system studies. In doing so, we will be using some simple, static models which can, in fact, be solved analytically. This will allow us to concentrate on the study technique. In each case, a realistic study would require a more complex, dynamic model. The precise, simple relationships used here would mostly be replaced by empirical relationships derived from many simulation runs.

2-8

System Analysis

To demonstrate a system analysis study, we will consider a small part of the corporate model developed earlier. Looking at just the financial and the production models, and ignoring the influence of the economies that might affect results, gives the simple system illustrated in Fig. 2-6. The system involves four variables denoted by the following symbols:

K Capital investment
L Labor
M Machinery
S Supply

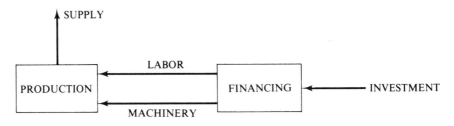

Figure 2-6. Portion of corporate model.

With regard to production, economists have found that the output of an enterprise is often related to the investment in labor and machinery by an equation of the general form

$$S = fL^{a_1}M^{a_2}$$

where f is a constant. Models of this form are known as *Cobb-Douglas models*, (3). They imply that a given percentage increase in the applied resources produces a proportional percentage increase in output. For simplicity, we will assume here that the exponents, a_1 and a_2, are both equal to one.[1] The production function then takes the form:

$$S = fLM$$

The financial model assumes that there are possible substitutions between labor and machinery. In the present case, we will assume a linear relationship, as follows:

$$K = eL + M$$

The coefficient e is a constant. It implies that one unit of investment in labor is equivalent to investing e units in machinery.

Note that both equations depend upon L and M. The financial model shows how a given amount of investment, K, can be divided between L and M. The production model shows how the output, or supply S, depends upon the amount invested in L and M. It should be possible to find an assignment of L and M that maximizes supply for a given investment.

The simple forms assumed for the equations permit differential calculus methods to be applied to show that there is, indeed, a maximum; but the result can also be demonstrated graphically, as shown in Fig. 2-7. The horizontal axis plots M, and the vertical axis plots L. The straight lines of Fig. 2-7 show the relationship between L and M for a given value of K, when it is assumed that the value of the coefficient e is 0.75. One line is for $K = 50$, and the other is for $K = 100$. As can be seen, increasing the value of K moves the line farther from the origin. The relationship between L and M for a constant value of S is not linear. It gives rise to the curves of Fig. 2-7, which are for the cases of $S = 50$ and 200 when f has the value 0.1. Again, larger values move the curve farther from the origin. The curve for $S = 50$ crosses the line for $K = 50$. This means that there are some combinations of values for L and M which produce a supply of 50 but add up to a K of less than 50. On the other hand, the curve for $S = 200$ does not cross the line for $K = 50$, meaning that it is impossible to produce a supply of 200 with an investment of only 50. The most that can be produced with $K = 50$ is given by the curve which just touches, or is tangent to, the line for $K = 50$. It happens to be the curve for $S = 83.3$. The maximum occurs when $L = 33.3$ and $M = 25$.

---

[1] In the terms of economic theory, the exponents are the elasticities of production with respect to labor and machinery investments. The special case of elasticities of one means that the percentage change in output equals the percentage change in either of the other two variables: a situation that would make an investor indifferent to making any change. See (12).

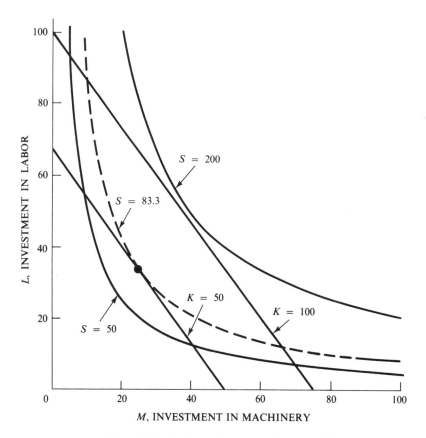

Figure 2-7. Finding optimum allocation of capital.

The model that has been analyzed is a small part of the full corporate model. An analysis of the full model would attempt to optimize some broader measure of the system performance, for instance, the rate of return on investment, which is the ratio of the profit to total capital investment. Such an analysis would, of course, mean extending the set of equations to cover the other segments of the full model. It would not necessarily invalidate the analysis of the submodel that has just been carried out.

What we have done is typical of many system analysis studies: we have sub-optimized a portion of a complete model. *Suboptimization* is a convenient way of breaking the optimization of a total system model into a number of simpler (approximate) steps. It is assumed that when the system as a whole is operating at its maximum, its various parts are also operating at *their* maxima. The system maximum can, therefore, be approached by maximizing the parts independently. This is not usually true in the strict mathematical sense. In the present case, for example, it does not follow that the best rate of return, taking into account all parts of the system, is achieved .when the supply is the maximum that can be achieved with the given investment. Realistically, it is unlikely that, when the rate of return is maximized, the supply is much different from that achieved with suboptimization; however, the uncertainty is the price paid for simplifying the solution.

Corporate models, particularly when used for the type of study just carried out, often are static models, like the one we have used. The models, however, will often be dynamic. When growth is being considered, and assumptions about such factors as competition and the state of the economy are included in the study, dynamic models are needed. Simulation methods are then needed for deriving the results on which the analysis is based.

2-9

System Design

To illustrate a system design problem, we consider the problem of designing a small part of an on-line computer system, that is, a computer that responds immediately to messages it receives. The function of the system is described by the flowchart of Fig. 2-8. Messages are being received over a communication channel at the rate of M messages a second. On the average, there are m characters in a message. They are received in a buffer that can hold a maximum of b characters. A fraction k of the messages need replies which have an average of r characters. The same buffer is used for both receiving and sending messages.

Processing a message when it has been received takes about 2,000 instructions. Preparing a reply requires a program that uses about 10,000 instructions. The finite size of the buffer means that messages and replies must be broken into sections of not more than b characters. Servicing each such section causes an interruption of the processing, requiring the execution of 1,000 instructions to transfer data either in or out of the computer.

Three computers are being considered as possible system components: a slow, a medium, and a fast computer, which have instruction execution rates of 25,000, 50,000, and 100,000 instructions a second. In addition, four buffer sizes are being considered: there could be a one-, two-, five-, or ten-character buffer. The design problem is to determine which of the possible twelve combinations of computer speed and buffer size will be capable of carrying out the processing. Given prices, the cheapest design can then be determined.

The crux of the design problem is to see that the computer can keep up with the flow of data. We must calculate how many characters per second are transferred in and out of the computer, and compare this with the number of instructions that

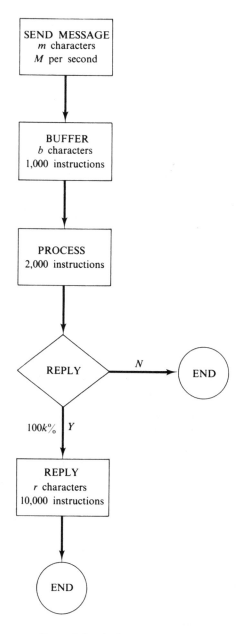

Figure 2-8. An input-output system.

have to be executed every second. We need not be concerned with the exact sequence of events, as the use of the buffer is switched between these functions; we need only consider the total number of events involved. There are M messages coming into the computer and kM replies going out every second. This requires that $Mm + kMr$ characters pass through the buffer every second. Since the buffer holds b characters, there will be $M(m + kr)/b$ interruptions every second. Adding together the instructions needed to process messages, prepare replies, and service buffer interruptions, the number of instructions to be executed every second, denoted by N, is given by,

$$N = 2,000M + 10,000Mk + \frac{1,000M(m + kr)}{b}$$

If we denote the number of instructions per second that the computer can execute by s, the condition to be met for the computer to keep up with the data flow is that $N \leq s$.

To simplify the discussion, suppose the details of the message traffic are specified with the following values:

$$M = 5 \qquad m = 15$$
$$k = 0.1 \qquad r = 50$$

Only the buffer size, b, and the computer speed, s, remain to be selected. Putting the given values into the formula for N, the condition, after simplification, can be written

$$\frac{20}{b} \leq \frac{s}{5,000} - 3$$

By trial, we can find which combinations of values satisfy the inequality. Alternatively, the solutions can be found by plotting. Figure 2-9 plots $20/b$ against b on log-log paper. The value of $20/b$ is plotted on the vertical scale, and the value of b, on the horizontal scale. The result is the diagonal line, sloping downwards to the right. The dotted vertical lines are drawn at the points 1, 2, 5, and 10, which are the possible buffer sizes. The horizontal dotted lines plot $s/5,000 - 3$ for the three values of computer speed. The twelve points of intersection of the vertical and horizontal lines represent the possible combinations of system components. The inequality, representing the conditions to be met, corresponds to saying the intersection must lie on or above the diagonal line. It can be seen that three combinations are satisfactory: the fast computer can operate with a two-character buffer, the medium-speed computer with a five-character buffer, and the low-speed computer will just keep up with a ten-character buffer. Realistic cost data will almost certainly result in the low-speed computer being the cheapest feasible design.

The problem in this case has been simplified to the point where a simple hand calculation solves the problem; further, the problem has been expressed in a form

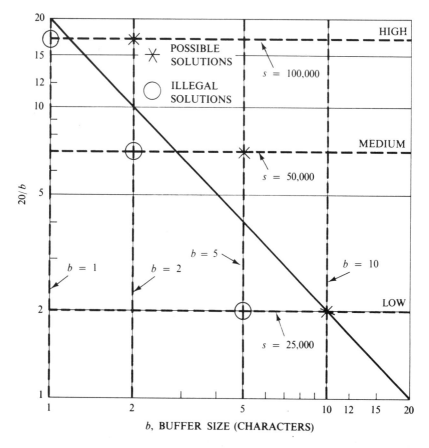

Figure 2-9. Solutions of the input-output model.

that allows a static model to be used. In practice, there will be mixtures of message types, with variations in sizes, even within a type; there will be congestion for getting access to data for generating replies; there will be competition for storage space to hold programs—all leading to delays in producing a reply. A simulation study would be needed to treat the more realistic problem.

2-10

System Postulation

As a last example in this chapter, we illustrate the use of system postulation by looking at a study designed to investigate the function of the liver in the human body, (5). When a chemical, thyroxine, is injected into the blood stream, it is carried to the liver. The liver can change thyroxine into iodine which is absorbed into the bile. However, neither the conversion nor the absorption occur instantaneously. Some of the thyroxine reenters the blood stream to be recirculated and returned to the liver. By using radioactive isotopes, it is possible to measure the rate at which thyroxine is removed from the blood stream, but the precise mechanism by which it is transferred from the blood to the liver and then to the bile was not known.

In the study, a mathematical model was constructed assuming that the body can be represented as three compartments and that the rates at which thyroxine is transferred between the compartments are proportional to the concentration of thyroxine in the compartments. Figure 2-10, reproduced from Ref. (5), illustrates the model and shows the assumed transfer coefficients between compartments. The compartments 1, 2, and 3 represent the blood vessels, the liver, and the bile, respectively. The model leads to three simple differential equations which are shown, together with their general solutions, in Fig. 2-10.

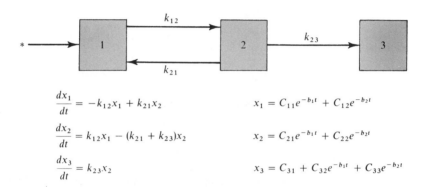

$$\frac{dx_1}{dt} = -k_{12}x_1 + k_{21}x_2 \qquad\qquad x_1 = C_{11}e^{-b_1 t} + C_{12}e^{-b_2 t}$$

$$\frac{dx_2}{dt} = k_{12}x_1 - (k_{21} + k_{23})x_2 \qquad\qquad x_2 = C_{21}e^{-b_1 t} + C_{22}e^{-b_2 t}$$

$$\frac{dx_3}{dt} = k_{23}x_2 \qquad\qquad x_3 = C_{31} + C_{32}e^{-b_1 t} + C_{33}e^{-b_2 t}$$

Figure 2-10. Mathematical model of the liver.

Figure 2-11, also reproduced from Ref. (5), shows the results of comparing actual measurements against the model predictions for one set of assumed values of the coefficients. It can be seen that the match between the results of the theoretical model and the experiment is close, which suggests that the hypothesis on transfer rates is reasonably accurate and allows estimates of the transfer coefficients to be made. In fact, the particular study from which this example is quoted found a more comprehensive model fitted the experimental facts more closely. See also (2) and (11).

Models of the type just discussed are called *compartment models,* (4). They are used extensively in medical, biological (6), and ecological studies, (15). As the present example demonstrates, the models are essentially dynamic models, and, although this example was able to use hand calculations, the study of compartment models is greatly enhanced by using simulation methods.

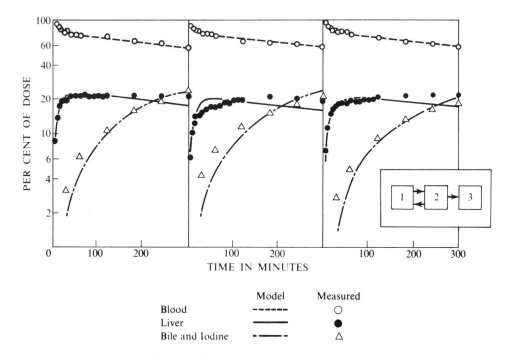

Figure 2-11. Results from model of the liver.

Bibliography

1 AMSTUTZ, ARNOLD E., *Computer Simulation of Competitive Market Response*, Cambridge, Mass.: The MIT Press, 1967.

2 BERGMAN, RICHARD N., AND MAHMOUD EL REFAI, "Dynamic Control of Hepatic Glucose Metabolism—Studies by Experiment and Computer Simulation," *Ann. Biomed. Eng.*, III, no. 4 (1975), 411–432.

3 COBB, CHARLES W., AND PAUL H. DOUGLAS, "A Theory of Production," *Amer. Econ. Rev.*, Supplement, (1925), 139–165.

4 EVERT, C. F., AND M. F. RANDALL, "Formulation and Computation of Compartment Models," *J. Pharm. Sci.*, LIX, no. 3 (1970), 102–114.

5 HAZELRIG, JANE B., "The Impact of High-Speed Automated Computation on Mathematical Models," *Mayo Clin. Proc.*, XXXIV, no. 11 (1964), 841–848.

6 JACQUEZ, JOHN A., *Compartment Analysis in Biology and Medicine*, New York: Elsevier Scientific Publishing Company, 1972.

7 MIHRAM, G. ARTHUR, "The Modeling Process," *IEEE Trans. Syst. Man Cybern.*, vol. SMC-2, no. 5 (1972), 621–629.

8 NAYLOR, THOMAS H., AND CHARLES JEFFRESS, "Corporate Simulation Models: A Survey," *Simulation*, XXIV, no. 6 (1975), 171–176.

9 PACKER, ARNOLD H., *Models of Economic Systems: A Theory for Their Development and Use*, Cambridge, Mass.: The MIT Press, 1972.

10 PITCHFORD, JOHN DAVID, *Population in Economic Growth*, Amsterdam: North-Holland Publishing Co., 1974.

11 RODECAP, SHARON E., AND F. T. LINDSTROM, "Numerical Simulation of a Compartmental System for the Distribution of Lipid Soluable Chemicals in Mammalian Tissues," *Comput. Biol. Med.*, VI, no. 1 (1976), 33–51.

12 SAMUELSON, PAUL A., *Economics*, (10th ed.), chap. 20, New York: McGraw-Hill Book Company, 1976.[2]

13 SCHRIEBER, ALBERT N., ed., *Corporate Simulation Models*, Seattle, Wash.: University of Washington Press, 1970.

14 THEIL, H., *Applied Economic Forecasting*, Amsterdam: North-Holland Publishing Co., 1975.

15 WEBB, WARREN L., HENRY J. SCHROEDER, Jr., AND LOGAN A. NORRIS, "Pesticide Residue Dynamics in a Forest Ecosystem: A Compartment Model," *Simulation*, XXIV, no. 6 (1975), 161–169.

[2]See footnote 1, this chapter.

3

SYSTEM SIMULATION[1]

3-1

The Technique of Simulation

Given a mathematical model of a system, it is sometimes possible to derive information about the system by analytical means. Where this is not possible, it is necessary to use numerical computation methods for solving the equations. The factor that distinguishes analytical methods from numerical methods is that analytical methods directly produce general solutions—numerical methods produce solutions in steps; each step gives the solution for one set of conditions, and the calculation must be repeated to expand the range of the solution.

Sometimes the term "simulation" is used to describe any procedure of establishing a model and deriving a solution numerically. However, in the case of static models, such as the corporate model, there seems to be little point in using the term "simulation" in preference to the general term "numerical computation," since no particular method of computation is being distinguished. In the case of dynamic models, however, such a distinction can be made.

Dynamic models can be solved analytically, as was seen in the case of the automobile wheel model discussed in Sec. 1-10. If the model needs to be solved numer-

---

[1]Parts of this chapter are based on Chap. 1 of (6). Figures 3-2 and 3-6 are adapted from the same.

ically, the particular technique that has come to be called simulation is the process of solving the equations of the model, step by step, with increasing values of time. As a result, the current values at any step of the computation represent the state of the system being modeled at that point in time. We will be seeing some examples shortly. Before doing so, however, let's look at an example of solving a dynamic model numerically, which would *not* be considered simulation, using this point of view.

Although the analytic solution to the automobile wheel model was not given in Sec. 1-10, it was mentioned that the condition for the motion of the wheel not to oscillate is that $\zeta \geq 1$. Without going into details, this fact can be derived from knowing what values of p solve the following auxiliary equation, (9):

$$p^2 + 2\zeta\omega p + \omega^2 = 0$$

This equation is related to the equation of motion describing the system (which is Eq. 1-1 of Sec. 1-10) by replacing the first and second derivatives of x with p and p^2, respectively, and setting the value of x equal to 1. Because this equation is of the second order it can be solved analytically. A more complex model, however, might lead to a much higher order auxiliary equation. The nature of the response can still be determined by finding the solutions to the auxiliary equation but it may be impossible to find the solutions analytically: instead, the equation must be solved numerically, (11).

The information derived numerically in this way does not constitute simulation as it will be defined here: it does not involve knowing the actual motion of the system. If the information were derived by simulation, the motion of the system under various conditions *would* be determined, and any conclusions about the nature of the motion would be drawn from these results—just as we observed from Fig. 1-9 that, for the second-order system, oscillations appear to occur only when ζ is less than 1.

To distinguish this interpretation of simulation from more general uses of the term we will use the term "system simulation." We therefore define *system simulation* as the technique of solving problems by the observation of the performance, over time, of a dynamic model of the system. This definition is broad enough to include the use of dynamic physical models, in which case the results are derived from physical measurements rather than numerical computations. It is not intended here to discuss further the use of physical models, therefore, future references to simulation will be in terms of mathematical models and numerical computations.

In some problems the time element is not significant, but a step-by-step calculation representing the successive stages of development in a system is appropriate. For example, one can evaluate the outcome of a sequential decision process, in which a number of successive choices must be made with given probabilities assigned to each choice, step by step, to find the probability of reaching each particular final outcome, (2). Problems of this nature are often solved by simulation programs, in which use is made of the intrinsic step-by-step processing built into the program. In such simulations, time is not explicitly involved. In effect, the model in use is assumed to change at uniform intervals of time and, since only the results of the final step are of interest, there is no purpose in giving a specific value to the interval; it is therefore set to zero.

3-2

The Monte Carlo Method

A particular numerical computation method, called the *Monte Carlo method*, consists of experimental sampling with random numbers, (8). For example, the integral of a single variable over a given range corresponds to finding the area under the graph representing the function. Suppose the function, $f(x)$ is positive and has lower and upper bounds a and b, respectively. Suppose, also, the function is bounded above by the value c. As shown in Fig. 3-1, the graph of the function is

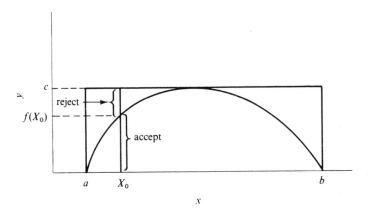

Figure 3-1. The Monte Carlo method.

then contained within a rectangle with sides of length c, and $b - a$.[2] If we pick points at random within the rectangle, and determine whether they lie beneath the curve or not, it is apparent that, providing the distribution of selected points is uniformly spread over the rectangle, the fraction of points falling on or below the

[2]It is not required that the function take the value of the upper bound. If it does not, however, the calculation is less efficient. We assume the bound is taken, to simplify the discussion.

curve should be approximately the ratio of the area under the curve to the area of the rectangle. If N points are used and n of them fall under the curve, then, approximately,

$$\frac{n}{N} = \int_a^b \frac{f(x)\,dx}{c(b-a)}$$

The accuracy improves as the number N increases.[3] When it is decided that sufficient points have been taken, the value of the integral is estimated by multiplying n/N by the area of the rectangle, $c(b-a)$.

The computational technique is illustrated in Fig. 3-1. For each point, a value of x is selected at random between a and b, say X_0. A second random selection is made between 0 and c to give Y. If $Y \le f(X_0)$ the point is accepted in the count n, otherwise it is rejected and the next point is picked.

It is not likely that the Monte Carlo method would be used to evaluate an integral of a single variable: there are more efficient numerical computation methods for that purpose. However, the method is often used on integrals of many variables by using a random number for each of the variables, (16). It will be noticed that, although random numbers have been used, the problem that has been solved is essentially determinate. There are many other applications of the Monte Carlo method, including examples where the problem being solved is of a statistical nature, such as in calculations concerned with the reliability of nuclear reactors, (5).

Monte Carlo applications are sometimes classified as being simulations. In addition, simulation is sometimes described as being an application of the Monte Carlo method, presumably because so many simulations involve the use of random numbers. Simulation and Monte Carlo are both numerical computational techniques. Simulation, as it will be described in this text, applies to dynamic models. The Monte Carlo technique is a computational technique applied to static models.

3-3

Comparison of Simulation and Analytical Methods

Compared with analytical solution of problems, the main drawback of simulation is apparent: it gives specific solutions rather than general solutions. In the study of automobile wheel motion, for example, an analytical solution gives all the

[3]Originally, the term Monte Carlo implied the use of some variance-reduction technique to reduce the number of trials. With the advent of digital computers, the savings achieved are less important. As a result, the term has generally come to mean any form of statistical experimentation.

conditions that can cause oscillation. Each execution of a simulation tells only whether a particular set of conditions did or did not cause oscillation. To try to find all such conditions requires that the simulation be repeated under many different conditions. The mathematical solution is obviously preferable, particularly when the solution being sought is some maximizing condition. A single mathematical solution might give such a condition, whereas many simulation runs may be needed to find a maximum, and yet still leave undecided the question of whether it is a local or global maximum.

The range of problems that can be solved mathematically is limited, however. Mathematical techniques require that the model be expressed in some particular format—for example, that it be in the form of linear algebraic equations or continuous linear differential equations. The system must be approximated or abstracted in order to derive a model that fits the format of a mathematical tecnique. All too often, the process amounts to making the problem fit the solution rather than the other way round.

In approaching a system study, it is essential to consider first what mathematical techniques might be applied to derive analytical solutions. It is a matter of judgment to decide whether the degree of abstraction required to apply analytical methods is too severe. To make that judgment it is necessary to consider carefully what questions need to be answered in the system study, and to what degree of accuracy the answers need to be known.

When the decision is made to simulate in order to use a more realistic model, it is still important to limit the amount of detail in the model to the minimum level necessary. The step-by-step nature of the simulation technique means that the amount of computation increases very rapidly as the amount of detail increases. Coupled with the need to make many runs to explore the range of conditions, the extra realism of simulation models can result in a very extensive amount of computing.

In many ways, the ideal way of using simulation is as an extension of mathematical solutions that might have been obtained at the cost of too much simplification. There are many simple limitations on a system, such as physical stops, finite time delays, or nonlinear forces, which render what would otherwise be a soluble mathematical model insoluble. Simulation easily removes these limitations, and it can then provide a powerful extension of known mathematical solutions.

Even when a model with a known solution is considered adequate, there are still times when a simulation will provide a quicker or more convenient way of deriving results. Many analytical results occur in the form of complex series or integrals that still require extensive evaluation. It is often more convenient to use simulation to obtain results directly from a model with specific values rather than to perform the numerical evaluation of the analytical solution, (1).

3-4
Experimental Nature of Simulation

The simulation technique makes no specific attempt to isolate the relationships between any particular variables; instead, it observes the way in which all variables of the model change with time. Relationships between the variables must be derived from these observations. A simulation study of the automobile wheel suspension system that was analyzed in Sec. 1-10 would proceed by following the motion of the wheel under different conditions. The relationship between D, K, and M to prevent oscillation, which was previously discovered analytically, would have to be discovered by observing the values that result in the motion being non-oscillatory. Simulation is, therefore, essentially an experimental problem-solving technique. Many simulation runs have to be made to understand the relationships involved in the system, so the use of simulation in a study must be planned as a series of experiments.

3-5
Types of System Simulation

In Sec. 1-4 we distinguished between continuous and discrete systems as being systems in which smooth or sudden changes occur. We went on, however, to comment that the important distinction, from the point of view of a system analyst, is whether the *model* selected to represent the system is continuous or discrete, since there is not a unique correspondence between types of systems and models.

The distinction between continuous and discrete models, however, was not made in the classification of models shown in Fig. 1-5, because this distinction does not determine whether analytical or numerical techniques will be applied to the model, a fact which *was* made a point of distinction. The distinction between continuous and discrete models becomes important when it is decided to use simulation, particularly when the simulation is to be carried out on a computer and a programming system is to be selected to perform the task. The general computational techniques used with the two kinds of model differ significantly. We will be discussing continuous and discrete simulation methods more fully in later chapters. Before doing so, however, we will show, in the next two sections, the general nature of the computational techniques that are used.

3-6
Numerical Computation Technique
for Continuous Models

To illustrate the general numerical technique of simulation based on a continuous model, consider the following example, (6). A builder observes that the *rate* at which he can sell houses depends directly upon the number of families who do not yet have a house. As the number of people without houses diminishes, the rate at which he sells houses drops. Let H be the potential number of households, and y be the number of families with houses. The situation is represented in Fig. 3-2;

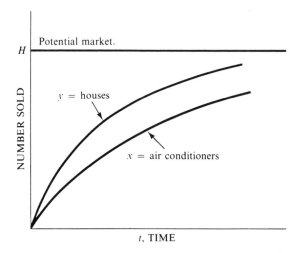

Figure 3-2.   Sale of houses and air conditioners.

the horizontal line at H is the total potential market for houses. The curve for y indicates how the number of houses sold increases with time. The slope of the curve (i.e., the rate at which y increases) decreases as $H - y$ gets less. This reflects the slowdown of sales as the market becomes saturated. Mathematically, the trend can be expressed by the equation

$$\dot{y} = k_1(H - y), \qquad y = 0 \text{ at } t = 0$$

Consider now a manufacturer of central air conditioners designed for houses. His rate of sales depends upon the number of houses built. (For simplicity, it is assumed that all houses will install an air conditioner.) As with house sales, the rate of sales diminishes as the unfilled market diminishes.

Let x be the number of installed air conditioners. Then the unfilled market is the difference between the number of houses and the number of installed air conditioners. The sales trend may be expressed mathematically by the equation

$$\dot{x} = k_2(y - x), \qquad x = 0 \text{ at } t = 0$$

The change of x with time is also illustrated in Fig. 3-2. The two equations constitute a model of the growth of air conditioner sales. Because of its simplicity, it is in fact possible to solve the model analytically. However, it quickly becomes insoluble if it is expanded to become more representative of actual marketing conditions. The market limit, for example, may not be stable. It could grow with population growth or fluctuate with economic conditions. The coefficients that determine the rates of growth could be influenced by the amount of money spent on advertising, and there could be competitive influences, such as mobile homes or apartment housing. These influences could also depend upon the population growth or prevailing economic conditions, and so further complicate the model.

The simple model, however, will serve to illustrate the general methods applied in continuous simulation. The simulation technique is to compute the output, step by step. Suppose that the computation is made at uniform intervals of time and that the calculation has already progressed to the time t_i, when the two variables of the problem have the values y_i and x_i. Figure 3-3 shows the next step in the calculation.

The calculation steps forward an interval Δt to $t_{i+1} = t_i + \Delta t$. The rates of sales are assumed to be constant over the interval. The rates can be interpreted as

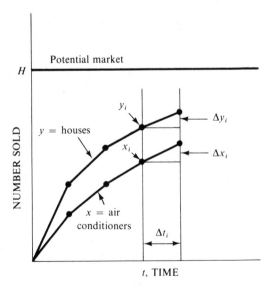

Figure 3-3. Calculations for air-conditioner sales model.

the amount of change per unit time. That is,

$$\text{rate of change of } y = \frac{\Delta y_i}{\Delta t}$$

$$\text{rate of change of } x = \frac{\Delta x_i}{\Delta t}$$

From the equations of the model, these may be written

$$\Delta y_i = k_1(H - y_i)\,\Delta t$$
$$\Delta x_i = k_2(y_i - x_i)\,\Delta t$$

Since y_i and x_i are known, it is a simple matter to get the values of y and x at time t_{i+1}. However, it will be noticed that the equation for Δy_i must be solved first to get the value of y_i needed in the equation for x_i. In preparation for the solution of a continuous system model, therefore, there must be a careful sorting of the equations to establish a workable order.

Repetition of the calculation using the new values of y and x produces the output at the end of the next interval. As illustrated in Fig. 3-3, the calculation is equivalent to calculating the slope at each point and projecting a short straight line at that slope. The simulation output is a series of such line segments, approximating the continuous curve that represents the true output of the model.

The method described is a very simple way of integrating differential equations numerically, but it is not a very accurate method, unless small steps are used, compared with the rate at which the variables change. There are other much more accurate, and often more efficient, ways of integrating numerically which do not rely simply upon the last-known value of the variables, (13). Rather, they use several previous values to predict the rate at which the variables are changing. (Special methods are used to supply initial values to start the process.) In addition, the computation interval is often adjusted in size to match the rate at which the variables are changing.

There are many programming systems available that incorporate continuous system simulation languages. They usually include a number of computational methods for the user's selection.

3-7

Numerical Computation Technique
for Discrete Models

To illustrate the general computational technique of simulation with discrete models, consider the following example. A clerk begins his day's work with a pile of documents to be processed. The time taken to process them varies. He works through the pile, beginning each document as soon as he finishes the previous one, except that he takes a five minute break if, at the time he finishes a document, it is an hour or more since he began work or since he last had a break. We assume the times to process the documents are given. We will also keep a count of how many documents are left. The count will be intially set to the number of documents at the beginning of the day, and, in this simple model, we will assume that no documents arrive during the day. The count will be decremented for each completed job, and work will stop if the count goes to zero.

The computations involved can be organized as shown in Table 3-1. The ith row corresponds to the ith document. The first column numbers the documents. There

TABLE 3-1 Simulation of Document Processing

Document Number i	Start Time t_b	Work Time t_w	Finish Time t_f	Cumul. Time t_c	Break Flag F	Number of Jobs N
1	0	45	45	45	0	57
2	45	16	61	61	1	56
3	66	5	71	5	0	55
4	71	29	100	34	0	54
5	100	33	133	67	1	53
6	138	25	163	25	0	52
7	163	21	184	46	0	51

are then four columns giving various times, measured in minutes from time zero. The second column gives the time the clerk begins to work on a document, denoted by $t_b(i)$. The third column gives the time required to work on the document, denoted by $t_w(i)$, and the fourth column gives the time each document is finished. The fifth column contains the cumulative time since work started or since the last break, measured at the time each job is completed. This is denoted by $t_c(i)$. There is a sixth column which contains a flag, denoted by F, that takes the value 1 if the clerk should take a break after the ith document, and the value 0 if he should not. The clerk works until there are no more documents, or the time he finishes a document goes beyond some time limit.

The computation proceeds row by row, and from left to right. The first row shows that work starts on the first document at time zero. The processing time is 45 minutes, so the job is finished at 45, with a cumulative time (in this case, since the start of work) of 45. This is not long enough for a break, so the flag is set to 0. The count, which was initialized to 57 jobs is dropped to 56.

Because the flag is zero and the count is not zero, the second document is begun at 45. It needs 16 minutes for processing, which leads to a cumulative time of 61, so the flag is set to 1 to indicate that a break should be taken. Because of the five minute break, the third document starts at 66. The computation continues in this manner until either, N, the count of documents to be processed, drops to zero, or the finish time, t_f, reaches some limit representing the end of the work period.

3-8
Distributed Lag Models

The computation for the model of the previous section was straightforward, but it can be imagined that the record-keeping quickly becomes burdensome as the model becomes larger or more complex. The aid of a computer and a suitable programming language then becomes important. There are some applications of simulation, however, where the simple technique illustrated in the previous section can be applied without difficulty, even for large models. If the events all occur synchronously, at fixed intervals of time, the computation remains simple.

Models that have the properties of changing only at fixed intervals of time, and of basing current values of the variables on other current values and values that occurred in previous intervals, are called *distributed lag models*, (7). They are used extensively in econometric studies where the uniform steps correspond to a time interval, such as a month or a year, over which some economic data are collected. As a rule, these models consist of linear, algebraic equations. They represent a continuous system, but one in which the data is only available at fixed points in time.

As an example, consider the following simple mathematical model of the national economy, (17).[4] Let,

C be consumption,
I be investment,
T be taxes,
G be government expenditure,
Y be national income.

$$C = 20 + 0.7(Y - T)$$
$$I = 2 + 0.1\,Y$$
$$T = 0.2\,Y$$
$$Y = C + I + G$$

All quantities are expressed in billions of dollars.

This is a static mathematical model, but it can be made dynamic by picking a fixed time interval, say one year, and expressing the current values of the variables

[4]This is a simple, expository model. The full model, developed in Ref. (17), has 22 equations.

in terms of values at previous intervals. Any variable that appears in the form of its current value and one or more previous intervals is said to be a *lagged variable*. Its value in a previous interval is denoted by attaching the suffix $-n$ to the variable, where n indicates the interval, with 1 denoting the previous interval, 2 denoting the one prior to that, and so on. In the present example we will only go back one interval.

The static model could be made dynamic by lagging all the variables, as follows:

$$I = 2 + 0.1Y_{-1}$$
$$T = 0.2Y_{-1}$$
$$Y = C_{-1} + I_{-1} + G_{-1}$$
$$C = 20 + 0.7(Y_{-1} - T_{-1})$$

Given an initial set of values for all variables, the values of the variables at the end of one year can be derived. Taking these values as the new values of the lagged variables, the values can then be derived at the end of the second year, and so on. There are only four equations in five unknowns, so one variable must be specified for each interval.

It is not necessary, however, to lag all the variables as has just been done. Suppose there is one equation that expresses a single current variable in terms of lagged variables only. When this equation is solved, a second equation can be solved that involves the current value just derived from the first equation plus any lagged variables. A third equation can then involve the current values derived from the first two equations plus any lagged variables, and so on. Using this principle, a slight rearrangement of the model just given makes it depend upon only one lagged variable. Substituting for T and C in the original equation for Y gives

$$Y = 45.45 + 2.27(I + G)$$

The set of equations can then be written in the form:

$$I = 2 + 0.1Y_{-1}$$
$$Y = 45.45 + 2.27(I + G)$$
$$T = 0.2Y$$
$$C = 20 + 0.7(Y - T)$$

The only lagged variable is Y, and to solve the model an initial value of Y_{-1} must be given. With the lagged variable, the current value of I can be derived from the first equation. Suppose the values of G are supplied for all intervals. The next equation then gives the current value of Y from the current value of I. The next equation similarly gives the current value of T from the current value of Y, and the last equation gives the current value of C from current values of both Y and T. Taking the current value of Y as the new value of the lagged variable Y_{-1}, the calculations can then be repeated for the next interval of time.

In practice, of course, the variables are not arbitrarily lagged to produce a dynamic model. Extensive statistical analysis is needed to establish that the dependencies do indeed exist, as well as to produce values for the coefficients in the equations, (12).

Although distributed lag models are conceptually simple, and they can be computed by hand, computers are extensively used to run them. However, the value of the computer is its more conventional data-processing capability. Unlike other types of model used for simulation purposes, there is no need for a special programming language to organize the simulation task.

Econometric models of this nature have been built for the national economies of most of the major industrial countries. Many large corporations also use models reflecting the effects of the national economy on their industry, (3). Among the better known models of the U.S. economy are the models developed at the Wharton School of the University of Pennsylvania, (14), Professor Klein's models at the University of Chicago, (10), and a model at the Brookings Institute in Washington, D.C., (4).

3-9

Cobweb Models

A particularly simple, but nevertheless useful, distributed lag model can be constructed from the static market model that was given in Sec. 1-9. This related supply, S, and demand, D, to market price, P. To be more realistic, the supply should be made dependent upon the price from the previous marketing period, since that is the only figure available to the supplier at the time of making future plans. The demand, however, will respond to current price. Again assuming the market is cleared, the market model in distributed lag form is as follows:

$$Q = a - bP$$
$$S = c + dP_{-1}$$
$$Q = S$$

Given an initial value of price, P_0, the value of S at the end of the first interval can be derived. This determines the value of Q, since the market is cleared, and from this the new value of P can be derived. This becomes the value of P_{-1} used to calculate the values for the second interval, and so on. Figure 3-4 shows the fluctuations of price for the following two cases:

(a)   (b)

$P_0 = 1.0$ $P_0 = 5.0$

$a = 12.4$ $a = 10.0$

$b = 1.2$ $d = 0.9$

$c = 1.0$ $c = -2.4$

$d = 0.9$ $d = 1.2$

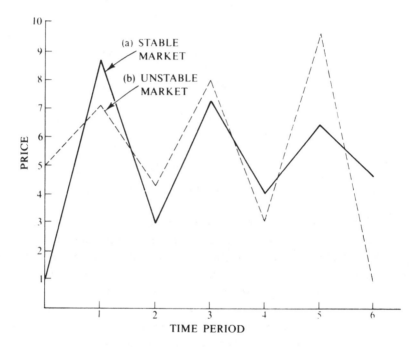

Figure 3-4. Fluctuations of market price.

As can be seen, case (a) represents a stable market in which the market settles to a price of 5.43 units. Case (b) is unstable, with price fluctuacting with increasing amplitude.

Models of this type are called *cobweb models* because of a particular way they can be solved graphically, (15). The method is illustrated in Fig. 3-5 for the stable case just considered. Two straight lines plot the linear relationships representing supply and demand, in the same way as they were plotted in Fig. 1-7 for the static market model. Beginning from the initial price of 1, a horizontal line to the supply

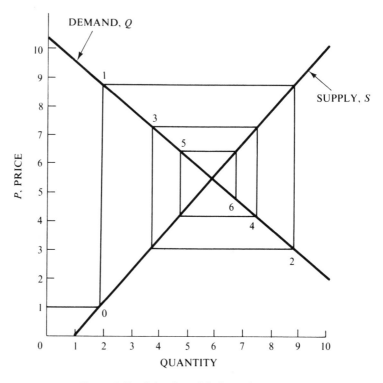

Figure 3-5. Cobweb model of a market economy.

line determines the supply that would be produced at that price level: about 1.8 units. A vertical line to the demand line reflects the fact that the market is cleared by making the demand equal the supply. With this small supply the price immediately rises to about 8.7 units. Taking that price as a basis for production, the supply in the next period jumps to about 8.8 units, as shown by the horizontal line, but, at this price, the supply is too big, therefore the price must drop to about 3.0 units in order to sell that amount. Continuing in this manner, each successive half-turn of the spiralling pattern produces the price in successive time periods. As can be seen, the spiral will eventually reach the equilibrium price of 5.43, from which time the market will remain unchanged.

3-10

Progress of a Simulation Study

It is apparent from the general discussion of the preceding sections that simulation is a very general method of studying problems. No formal procedure can be given for showing how a simulation study will proceed. There is not even a simple way of deciding whether to simulate or not, or, if simulation is used, whether to use a continuous or discrete model. However, some of the steps involved in the progress of a simulation study are illustrated by the flowchart of Fig. 3-6. The flowchart is illustrative: it is not intended to suggest a formal procedure.

An initial step is to describe the problem to be solved in as concise a manner as possible so that there is a clear statement of what questions are being asked and what measurements need to be taken in order to answer those questions. Based on this problem definition, a model must be defined. It is at this point that it becomes apparent whether the model can be kept in a form that allows analytical techniques to be used.

It should be understood that there is not, in fact, a single model for any given system. In the course of a study *many* different models are likely to be constructed as understanding of the system behavior increases. A possible course, which is to be recommended if it does not entail too much extra effort, is to explore first a model that can be solved analytically, even though it is clear that the simplifications required to produce the model are too drastic. The results will help guide the simulation study.

When it is decided to simulate, the experimental nature of the simulation technique makes it essential to plan the study by deciding upon the major parameters to be varied, the number of cases to be conducted, and the order in which runs are to be made. This procedure will help gauge the magnitude of the simulation effort and may cause a reappraisal of the model. Of course, it is not possible to plan every step ahead of time, and the study plan may need to be revised periodically as results become available; but the plan should always be able to show the direction in which the study is going.

Given that the simulation is to be on a digital computer, a program must be written. Actually this step is likely to be carried out in parallel with the study planning, once the model structure has been decided. Figure 3-6 does not show the choice between using a continuous or discrete model. The figure does show, however, the step of establishing the validity of the model before beginning the major set of runs. This is an important step that requires good judgment. It is worth bearing in mind the need for validation at the time of deciding upon the model. It could be that some of the earlier, oversimplified models may provide guidelines for establishing the reasonableness of the final model.

The study will then move into the stage of executing a series of runs according to the study plan. As has already been intimated, the study is not likely to proceed in the orderly steps implied by the flowchart of Fig. 3-6. As results are obtained, it is quite likely that there will be many changes in the model and the study plan. Frequently, the main value of the early runs in a simulation study is to gain insight into the general behavior of the system, rather than to extract data. In the construction of a model certain parameters of the system are clearly going to be important

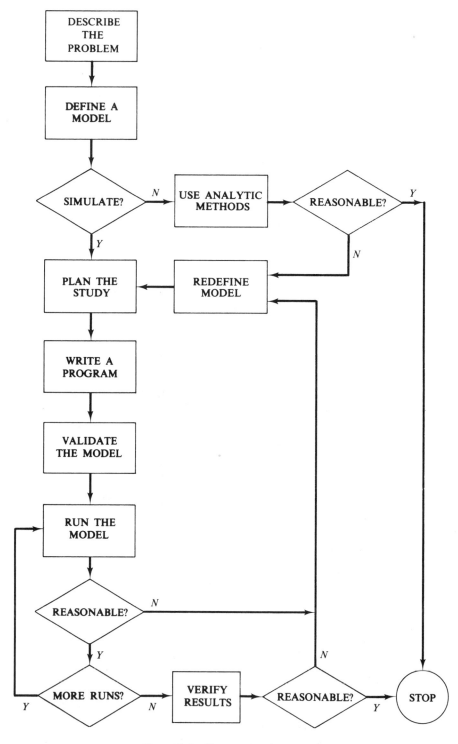

Figure 3-6. The process of simulating.

and others unimportant—to the point where they can be neglected or simplified. There will also be parameters whose importance, either by themselves or in interaction with other parameters, is not so clear. The early runs may make their significance clear, and so lead to a reassessment of the model. This is implied by the blocks in Fig. 3-6 that ask the question of whether the results are reasonable or not. This should not, however, be interpreted as an incentive to include numerous parameters in the model on the assumption that the unimportant ones can be quickly eliminated. As was emphasized in Sec. 3-4 it is essential to keep the model as simple as possible.

A major factor that needs to be considered in the presence of random variables is the problem of statistically verifying the results. The flowchart of Fig. 3-6 shows this step as following the runs, but, in fact, it must also be considered on each planned run. Once random variables enter a model, almost all measures of interest will also be random variables. The results of a single run will give just one sample of the measured variable. It is thus essential to repeat the run with a different set of random numbers so that more than one sample is available; it is then possible to judge the variability of the results. The number of repetitions needed depends upon the nature of the system and the importance to be attached to the results.

Sometimes it is useful to repeat runs so that parts of the model have different random numbers while the rest use exactly the same random numbers on each run. By comparing the results obtained under these conditions with the case where all parts of the system have different random numbers, it is then possible to analyze the cause of the variance in the results. The problem of verifying simulation results will be discussed more fully in Chap. 14.

Exercises

3-1 Draw the cobweb model of Sec. 3-9 for the unstable conditions (case b).

3-2 Draw a cobweb model for the following market:

$$D = 12.4 - 1.2P$$
$$S = 8.0 - 0.6P_{-1}$$
$$P_0 = 1.0$$

3-3 Derive the mathematical conditions under which the market model of Sec. 3-9 is stable. Interpret the results graphically. (Hint: Express the equations in terms of the deviations of price and quantity from the equilibrium values.)

3-4 The supplier in the market model of Sec. 3-9 learns to discount the swings in market price by adjusting the price upon which he bases the supply. He computes the supply from the given curve but uses the price $P_{t-1} - r(P_{t-1} - P_{t-2})$.

Assume $P_0 = P_1 = 1.0$ and compute the market fluctuations. (The problem cannot be conveniently solved graphically. Use algebraic methods.)

3-5   Use a cobweb model to investigate a market in which the supply and demand functions are

$$D = \frac{17.91}{p^{1/2}} - 4.66$$
$$9S = 5.0(P_{-1} - 1)$$

Assume the market is always cleared.

3-6   Find the growth in national consumption for five years using the model given in Sec. 3-8. Assume the initial income Y_{-1} is 80 and take the government expenditure in the 5 years to be as follows:

Year	G
1	20
2	25
3	30
4	35
5	40

Bibliography

1   BAJAJ, K. S., "Simulation of Irrational and Transcendental Numbers," *Franklin Inst. J.*, CCC, no. 3 (1975), 193–196.

2   CHERNOFF, HERMAN, AND LINCOLN E. MOSES, *Elementary Decision Theory*, New York: John Wiley & Sons, Inc., 1959.

3   DAVIS, B. E., G. J. CACCAPPOLO, AND M. A. CHAUDRY, "An Econometric Planning Model for American Telephone and Telegraph Company," *Bell J. Econ. Manage. Sci.*, IV, no. 1 (1973), 29–56.

4   FROMM, G., AND P. J. TAUBMAN, eds., *The Brookings Model: Perspectives and Recent Developments*, Amsterdam: North-Holland Publishing Co., 1975.

5   GOOD, W., AND R. JOHNSTON, "A Monte Carlo Method for Criticality Problems," *Nucl. Sci. Eng.*, V, no. 5 (1959), 371–375.

6   GORDON, GEOFFREY, *The Application of GPSS V to Discrete System Simulation*, pp. 10–11, Englewood Cliffs, N. J.: Prentice-Hall, Inc., 1975.

7   GRILICHES, ZVI, "Distributed Lags: A Survey," *Econometrica*, XXXV, no. 1 (1967), 16–49.

8 HAMMERSLEY, J. M. AND D. C. HANDSCOMB, *Monte Carlo Methods*, London: Methuen & Company, 1964.

9 JAMES, HUBERT M., NATHANIEL B. NICHOLS, AND RALPH S. PHILIPS, eds., *Theory of Servomechanisms*, New York: Dover Publications, Inc., 1965.

10 KLEIN, LAWRENCE R., AND ARTHUR. S. GOLDBERGER, *An Econometric Model of the United States, 1929–1952*, Amsterdam: North-Holland Publishing Co., 1955.

11 LAGO, GLADWYN, AND L. M. BENNINGFIELD, *Control System Theory: Feedback Theory*, New York: The Ronald Press Company, 1962.

12 MALINVAUD, E., *Statistical Methods of Econometrics*, (2nd ed.), chap. 15, Amsterdam: North-Holland Publishing Co., 1970.

13 MARTENS, HINRICH R., "A Comparitive Study of Digital Integration Methods," *Simulation*, XII, no. 2 (1969), 87–96.

14 MCCARTHY, M. D., "The Wharton Quarterly Forecasting Model, Mark III," *Studies in Quantitive Economics Series*, (1969) no. 6. Philadelphia: Dept. of Economics, Wharton School of Finance and Commerce, University of Pennsylvania.

15 SAMULESON, PAUL A., *Economics*, (10th ed.), chap. 20, New York: McGraw-Hill Book Company, 1976.

16 SHREIDER, Y. A., *The Monte Carlo Method: The Method of Statistical Trials*, chap. 2, London: Pergamon Press, 1966.

17 SUITS, DANIEL, B., "Forecasting and Analysis with an Econometric Model," *Amer. Econ. Rev.*, LII, no. 2 (1962), 104–132.

4

CONTINUOUS
SYSTEM SIMULATION

4-1

Continuous System Models

A continuous system is one in which the predominant activities of the system cause smooth changes in the attributes of the system entities. When such a system is modeled mathematically, the variables of the model representing the attributes are controlled by continuous functions. The distributed lag model discussed in Sec. 3-8 is an example of a continuous model. That model described how the attributes of the system were related to each other in the form of linear algebraic equations. More generally, in continuous systems, the relationships describe the *rates* at which attributes change, so that the model consists of differential equations.

The simplest differential equation models have one or more linear differential equations with constant coefficients. It is then often possible to solve the model without the use of simulation. Even so, the labor involved may be so great that it is preferable to use simulation techniques. However, when nonlinearities are introduced into the model, it frequently becomes impossible or, at least, very difficult to solve the models. Simulation methods of solving the models do not change fundamentally when nonlinearities occur. The methods of applying simulation to continuous models can therefore be developed by showing their application to models where the differential equations are linear and have constant coefficients, and then generalizing to more complex equations.

4-2

Differential Equations

An example of a linear differential equation with constant coefficients was given in Sec. 1-8 to describe the wheel suspension system of an automobile. The equation derived was

$$M\ddot{x} + D\dot{x} + Kx = KF(t) \tag{4-1}$$

Note that the dependent variable x appears together with its first and second derivatives $\dot{x}$ and $\ddot{x}$, and that the terms involving these quantities are multiplied by constant coefficients and added. The quantity $F(t)$ is an input to the system, depending upon the independent variable t. A linear differential equation with constant coefficients is always of this form, although derivatives of any order may enter the equation. If the dependent variable or any of its derivatives appear in any other form, such as being raised to a power, or are combined in any other way—for example, by being multiplied together, the differential equation is said to be nonlinear.

When more than one independent variable occurs in a differential equation, the equation is said to be a *partial differential equation*. It can involve the derivatives of the same dependent variable with respect to each of the independent variables. An example is an equation describing the flow of heat in a three dimensional body. There are four independent variables, representing the three dimensions and time, and one dependent variable, representing temperature. The general method of solving such equations numerically is to use finite differences to convert the equations into a set of ordinary (that is, non-partial) differential equations, which can be solved by the methods that are about to be described. There are programming languages specially designed for this type of equation, which will perform the function of constructing the finite differences. See, for example, Ref. (6).

Differential equations, both linear and nonlinear, occur repeatedly in scientific and engineering studies. The reason for this prominence is that most physical and chemical processes involve rates of change, which require differential equations for their mathematical description. Since a differential coefficient can also represent a growth rate, continuous models can also be applied to problems of a social or economic nature where there is a need to understand the general effects of growth trends. We shall be discussing such socio-economic systems in the next chapter. For now, we concentrate on the solution of scientific and engineering types of problems which generally require a higher degree of accuracy than just establishing trends. This, however, is only a separation of types of application which have different emphases. The continuous system simulation methods that will be described in this chapter can be used for either type of application.

To illustrate how differential equations can represent engineering problems we will show how the equation describing the automobile wheel suspension system is derived from mechanical principles. If we pick a point of the wheel as a reference point from which to measure the vertical displacement of the wheel, the variable x can represent the displacement of the point, taking x to be positive for an upward movement. (See Fig. 1-6.) The velocity of the wheel, in the vertical direction, is the rate of change of position, which is the first differential, $\dot{x}$. The acceleration of the wheel, in the vertical direction, is the rate of change of the *velocity*, which is the second differential, $\ddot{x}$.

The mechanical law that determines the relationship between applied force and movement of a body states that the *acceleration* of the body is proportional to the force. In particular, if there is no force there is no acceleration. The body then remains stationary, if no force has been acting upon it, or it continues moving at a constant velocity—a fact that, at one time, would have seemed to be merely an abstraction but is familiar in this age of space travel, since the ideal condition of no force acting on a body is obtainable in space.

The coefficient of proportionality between the force and acceleration is the mass of the body; so, in the case of the automobile wheel, where the mass is M and the applied force is $KF(t)$, the equation of motion in the absence of other forces would be as follows:

$$M\ddot{x} = KF(t) \tag{4-2}$$

However, the shock absorber exerts a resisting force that depends on the velocity of the wheel: the force is zero when the wheel is at rest, and it increases as the velocity rises. If we assume the force is directly proportional to the velocity, it can be represented by $D\dot{x}$, where D is a measure of the viscosity of the shock absorber. Similarly, the spring exerts a resisting force which depends on the extent to which it has been compressed. (Assume that x is defined so that it is zero when the spring is uncompressed.) Again, if the force is directly proportional to the compression, it can be represented by Kx, where K is a constant defining the stiffness of the spring. Since both these forces oppose the motion, they subtract from the applied force to give the following equation of motion:

$$M\ddot{x} = KF(t) - D\dot{x} - Kx$$

Transposing terms, the equation is seen to be Eq. (4-1), which was previously given. It is a linear differential equation, with constant coefficients, and, as mentioned before, it can be solved analytically. A more accurate model, however, would not assume the restraining forces are linear functions of the motion variables, and so the coefficients would *not* be linear. Also, there are physical limits to the amount of movement that is possible, which places another nonlinearity on the equation. The more realistic model is unlikely to be soluble by analytic methods, so it would become the basis of a continuous system simulation study.

In Sec. 1-8 the equation that has just been derived was also used to describe an electrical system. We will not derive the equation for that interpretation, but the same general principles apply. The quantity being measured in that case is the electrical charge of the condenser. The physical laws of electricity relating current and voltage to charge introduce the first two derivatives of the charge, and so lead to a linear differential equation of the same form as that for the automobile wheel, but with different interpretations of the constant coefficients. Similar examples can be taken from many other fields of engineering.

4-3

Analog Computers

Historically, continuous system simulation was in general use for studying complex systems long before discrete system simulation was similarly applied. The main reason was that, before the general availability of digital computers, needed for the numerical computation of discrete system simulation, there existed devices whose behavior is equivalent to a mathematical operation such as addition or integration, (21). Putting together combinations of such devices in a manner specified by a mathematical model of a system allowed the system to be simulated. By their nature, the devices gave continuous outputs and so lent themselves readily to the simulation of continuous systems.[1] Specific devices have been created for particular systems but with so general a technique, it has been customary to refer to them as *analog computers*, or, when they are primarily used to solve differential equation models, as *differential analyzers*, (5). Physical models based on analogies were described in Chap. 1. The specific example discussed there described the analogy between electrical and mechanical systems. Simulation with an analog computer, however, is more properly described as being based on a mathematical model than as being a physical model.

The most widely used form of analog computer is the electronic analog computer, based on the use of high gain dc (direct current) amplifiers, called operational amplifiers, (17). Voltages in the computer are equated to mathematical variables, and the operational amplifiers can add and integrate the voltages. With appropriate

---

[1]Analog computers are sometimes used to solve static models. The result being sought might be a number that can be defined as the steady state solution of a differential equation system, in much the same way as an equilibrium market price can be determined by a cobweb model. (See Sec. 3-9.) For examples of this use of analog computers, see Ref. (7) and (20).

circuits, an amplifier can be made to add several input voltages, each representing a variable of the model, to produce a voltage representing the sum of the input variables. Different scale factors can be used on the inputs to represent coefficients of the model equations. Such amplifiers are called *summers*. Another circuit arrangement produces an *integrator* for which the output is the integral with respect to time of a single input voltage or the sum of several input voltages. All voltages can be positive or negative to correspond to the sign of the variable represented. To satisfy the equations of the model, it is sometimes necessary to use a *sign inverter*, which is an amplifier designed to cause the output to reverse the sign of the input.

Electronic analog computers are limited in accuracy for several reasons. It is difficult to carry the accuracy of measuring a voltage beyond a certain point. Secondly, a number of assumptions are made in deriving the relationships for operational amplifiers, none of which is strictly true; so, amplifiers do not solve the mathematical model with complete accuracy. A particularly troublesome assumption is that there should be zero output for zero input. Another type of difficulty is presented by the fact that the operational amplifiers have a limited dynamic range of output, so that scale factors must be introduced to keep within the range. As a consequence, it is difficult to maintain an accuracy better than 0.1 % in an electronic analog computer. Other forms of analog computers have similar problems and their accuracies are not significantly better.

A digital computer is not subject to the same type of inaccuracies. Virtually any degree of accuracy can be programmed and, with the use of floating-point representation of numbers, an extremely wide range of variations can be tolerated. Integration of variables is not a natural capability of a digital computer, as it is in an analog computer, so that integration must be carried out by numerical approximations. However, methods have been developed which can maintain a very high degree of accuracy.

A digital computer also has the advantage of being easily used for many different problems. An analog computer must usually be dedicated to one application at a time, although time-sharing sections of an analog computer has become possible.

In spite of the widespread availability of digital computers, many users prefer to use analog computers. There are several considerations involved. The analog representation of a system is often more natural in the sense that it directly reflects the structure of the system; thus simplifying both the setting-up of a simulation and the interpretation of the results. Under certain circumstances, an analog computer is faster than a digital computer, principally because it can be solving many equations in a truly simultaneous manner; whereas a digital computer can be working only on one equation at a time, giving the appearance of simultaneity by interlacing the equations. On the other hand, the possible disadvantages of analog computers, such as limited accuracy and the need to dedicate the computer to one problem, may not be significant.

4-4

Analog Methods

The general method by which analog computers are applied can be demonstrated using the second-order differential equation that has already been discussed:

$$M\ddot{x} + D\dot{x} + Kx = KF(t)$$

Solving the equation for the highest order derivative gives

$$M\ddot{x} = KF(t) - D\dot{x} - Kx$$

Suppose a variable representing the input $F(t)$ is supplied, and assume for the time being that there exist variables representing $-x$ and $-\dot{x}$. These three variables can be scaled and added with a summer to produce a voltage representing $M\ddot{x}$. Integrating this variable with a scale factor of $1/M$ produces $\dot{x}$. Changing the sign produces $-\dot{x}$, which supplies one of the variables initially assumed; and a further integration produces $-x$, which was the other assumed variable. For convenience, a further sign inverter is included to produce $+x$ as an output.

A block diagram to solve the problem in this manner is shown in Fig. 4-1. The symbols used in the figure are standard symbols for drawing block diagrams representing analog computer arrangements. The circles indicate scale factors applied to the variables. The triangular symbol at the left of the figure represents the operation of adding variables. The triangular symbol with a vertical bar represents an integration, and the one containing a minus sign is a sign changer.

$$M\ddot{x} + D\dot{x} + Kx = KF(t)$$

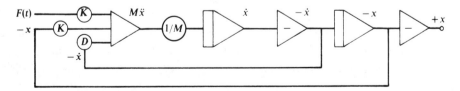

Figure 4-1. Diagram for the automobile suspension problem.

The addition on the left, with its associated scaling factors, corresponds to the addition of the variables representing the three forces on the wheel, producing a variable representing $M\ddot{x}$. The scale is changed to produce $\ddot{x}$, and the result is integrated twice to produce both $\dot{x}$ and x. Sign changers are introduced so that variables of the correct sign can be fed back to the adder, and the output can be given in convenient form.

With an electronic analog computer, the variables that have been described would be voltages, and the symbols would represent operational amplifiers arranged as adders, integrators, and sign changers. Figure 4-1 would then represent how the amplifiers are interconnected to solve the equation. It should be pointed out, however, that there can be several ways of drawing a diagram for a particular problem, depending upon which variables are of interest, and on the size of the scale factors.

When a model has more than one independent variable, a separate block diagram is drawn for each independent variable and, where necessary, interconnections are made between the diagrams. As an example, Fig. 4-2 shows a block dia-

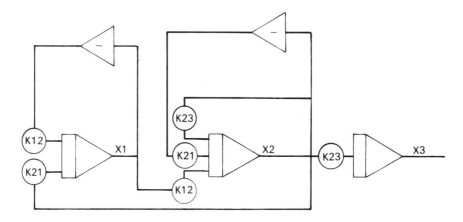

Figure 4-2. Analog computer model of the liver.

gram for solving the model of the liver shown in Fig. 2-10. There are three integrators, shown at the bottom of the figure. Reading from left to right, they solve the equations for x_1, x_2, and x_3. Interconnections between the three integrators, with sign changers where necessary, provide inputs that define the differential coefficients of the three variables. The first integrator, for example, is solving the equation

$$\dot{x}_1 = -k_{12}x_1 + k_{21}x_2$$

The second integrator is solving the equation

$$\dot{x}_2 = k_{12}x_1 - (k_{21} + k_{23})x_2$$

In this case, the variable x_2 is being used twice as an input to the integrator, so that the two coefficients k_{21} and k_{23} can be changed independently. The last integrator solves the equation

$$\dot{x}_3 = k_{23}x_2$$

4-5

Hybrid Computers

The scope of analog computers has been considerably extended by developments in solid-logic electronic devices, (1) and (11). Analog computers always used a few nonlinear elements, such as multipliers or function generators. Originally, such devices were expensive to make. Solid-logic devices, in addition to improving the design and performance of operational amplifiers, have made such nonlinear devices cheaper and easier to obtain. They have also extended the range of devices. Among the elements that can easily be associated with analog computers are circuits that carry out the logical operations of Boolean algebra, store values for later use, compare values, and operate switches for controlling runs.

The term *hybrid computer* has come to describe combinations of traditional analog-computer elements, giving smooth, continuous outputs, and elements carrying out such nonlinear, digital operations as storing values, switching, and performing logical operations. Originally, the term had the connotation of extending analog-computer capabilities, usually by the addition of special-purpose, and often specially constructed, devices. More recently, few purely analog computers are built. Instead, computers with large numbers of standard, nonlinear elements are readily available.

Hybrid computers may be used to simulate systems that are mainly continuous, but do, in fact, have some digital elements—for example, an artificial satellite for which both the continuous equations of motion, and the digital control signals must be simulated. The general technique of applying hybrid computers follows the methods outlined in Sec. 4-4, with blocks in the diagrams representing the non-linear elements as special functions, rather than as purely mathematical operators. Hybrid computers are also useful when a system that can be adequately represented by an analog computer model is the subject of a repetitive study. It might, for example, be necessary to search for a maximum value. Using digital devices to save values, and other devices that will change initial conditions, together with switches to control runs, a hybrid computer can be arranged to carry out large portions of the study without human intervention, (26).

4-6

Digital-Analog Simulators

To avoid the disadvantages of analog computers, many digital computer programming languages have been written to produce *digital-analog simulators*. They allow a continuous model to be programmed on a digital computer in essentially the same way as it is solved on an analog computer. The languages contain macro-instructions that carry out the action of adders, integrators, and sign changers. A program is written to link together these macro-instructions, in essentially the same manner as operational amplifiers are connected in analog computers.

More powerful techniques of applying digital computers to the simulation of continuous systems have been developed. As a result, digital-analog simulators are not now in extensive use. A history of their development will be found in Ref. (19).

4-7

Continuous System Simulation Languages
(CSSLs)

Confining a digital computer to routines that represent just the functions of an analog or a hybrid computer, as is done with a digital-analog simulator, is clearly a restriction. To remove the restriction, a number of *continuous system simulation languages* (abbreviated to CSSLs) have been developed. They use the familiar statement-type of input for digital computers, allowing a problem to be programmed directly from the equations of a mathematical model, rather than requiring the equations to be broken into functional elements, (8). Of course, a CSSL can easily include macros, or subroutines, that perform the function of specific analog elements, so that it is possible to incorporate the convenience of a digital-analog simulator. In fact, most implementations of a CSSL, in addition to including a subset of standard digital-analog elements, allow the user to define special purpose elements that correspond to operations that are particularly important in specific types of applications.

In extending beyond the provision of simple analog functions, such as addition and integration, CSSLs include a variety of algebraic and logical expressions to describe the relations between variables. They, therefore, remove the orientation toward linear differential equations which characterizes analog methods. A general specification of the requirements of a CSSL has been published by a user group, (23). Several implementations have been produced, (3). One particular CSSL that will be described here, to illustrate the nature of these languages, is the *Continuous System Modeling Program*, Version III, (CSMP III), (14), and (22).

4-8

CSMP III

A CSMP III program is constructed from three general types of statements:

Structural statements, which define the model. They consist of FORTRAN-like statements, and functional blocks designed for operations that frequently occur in a model definition.

Data statements, which assign numerical values to parameters, constants, and initial conditions.

Control statements, which specify options in the assembly and execution of the program, and the choice of output.

Structural statements can make use of the operations of addition, subtraction, multiplication, division, and exponentiation, using the same notation and rules as are used in FORTRAN. If, for example, the model includes the equation

$$X = \frac{6Y}{W} + (Z - 2)^2$$

the following statement would be used:

```
X = 6.0*Y/W + (Z - 2.0)**2.0
```

Note that real constants are specified in decimal notation. Exponent notation may also be used; for example, $1.2E-4$ represents 0.00012. Fixed value constants may also be declared. Variable names may have up to six characters.

There are many functional blocks which, in addition to providing operations specific to simulation, duplicate many of the mathematical-function subprograms of FORTRAN. Among these are the exponential function, trigonometric functions, and functions for taking maximum and minimum values. Figure 4-3 is a list of eleven of the functional blocks.

GENERAL FORM	FUNCTION
Y = INTGRL (IC, X) Y (0) = IC INTEGRATOR	$$Y = \int_0^t X dt + IC$$
Y = LIMIT (P$_1$, P$_2$, X) LIMITER	$Y = P_1 \quad\quad X < P_1$ $Y = P_2 \quad\quad X > P_2$ $Y = X \quad\quad P_1 \leqslant X \leqslant P_2$ 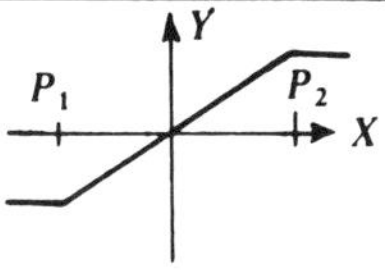
Y = STEP (P) STEP FUNCTION	$Y = 0 \quad\quad t < P$ $Y = 1 \quad\quad t \geqslant P$ 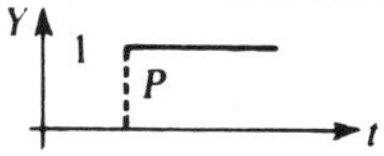
Y = EXP (X) EXPONENTIAL	$Y = e^X$
Y = ALOG (X) NATURAL LOGARITHM	$Y = \ln(X)$
Y = SIN (X) TRIGONOMETRIC SINE	$Y = \sin(X)$
Y = COS (X) TRIGONOMETRIC COSINE	$Y = \cos(X)$
Y = SQRT (X) SQUARE ROOT	$Y = X^{1/2}$
Y = ABS (X) ABSOLUTE VALUE (REAL ARGUMENT AND OUTPUT)	$Y = |X|$
Y = AMAX1 (X$_1$, X$_2$ ... X$_n$) LARGEST VALUE (REAL ARGUMENTS AND OUTPUT)	$Y = \max(X_1, X_2, \ldots, X_n)$
Y = AMIN1 (X$_1$, X$_2$ ... X$_n$) SMALLEST VALUE (REAL ARGUMENTS AND OUTPUT)	$Y = \min(X_1, X_2, \ldots, X_n)$

Figure 4-3. CSMP III functional blocks.

The following functional block is used for integration:

$$Y = INTGRL(IC,X)$$

where Y and X are the symbolic names of two variables and IC is a constant. The variable Y is the integral with respect to time of X, and it takes the initial value IC.

Of the data statements, one called INCON can be used to set the initial values of the integration-function block. Other parameters can be given a value for a specific run with the CONST control statement. A control statement called PARAM can also be used to assign values to individual parameters, but its chief purpose is to specify a series of values for one parameter (and only one). The model will be run with the specified parameter taking each of the values on successive runs of the same model. Examples of how these statements are written are

CONST A = 0.5, XDOT = 1.25, YDOT = 6.22
PARAM D = (0.25, 0.50, 0.75, 1.0)

As indicated, several values can be specified with each statement by separating the values with commas.

Among the control statements is one called TIMER, which must be present to specify certain time intervals. An integration interval size must be specified. For adequate accuracy, it should be small in relation to the rate at which variables change value. The total simulation time must also be given. Output can be in the form of printed tables and/or print-plotted graphs. Interval sizes for printing and plotting results need to be specified. The following is an example:

TIMER DELT = 0.005, FINTIM = 1.5, PRDEL = 0.1, OUTDEL = 0.1

The items specified are

DELT	Integration interval
FINTIM	Finish time
PRDEL	Interval at which to print results
OUTDEL	Interval at which to print-plot

If printed and/or print-plotted output is required, control statements with the words PRINT and PRTPLT are used, followed by the names of the variables to form the output. Several variables can be part of the output and they are listed with their names separated by commas. Two other control statements with the words TITLE and LABEL can be used to put headings on the printed and print-plotted outputs, respectively. Whatever comment is written after the words becomes the heading.

The set of structural, data, and control statements for a problem can be assembled in any order but they must end with an END control statement. However, control statements which define another run of the same model can follow an END statement if they also are terminated by another END statement. This can be repeated many times, until an ENDJOB statement signals end of all runs. A completely separate model can then follow the ENDJOB statement.

As an example, Fig. 4-4 shows a CSMP III program for the automobile wheel suspension problem. It has been coded for the case where $M = 2.0$, $F = 1$, and

```
TITLE AUTOMOBILE SUSPENSION SYSTEM
*
PARAM D = (5.656, 16.968, 39.592, 56.56, 113.12)
*
 X2DOT = (1.0/M)*(K*F - K*X - D*XDOT)
 XDOT = INTGRL(0.0,X2DOT)
 X = INTGRL(0.0,XDOT)
*
CONST M = 2.0, F = 1.0, K = 400.0
TIMER DELT = 0.005, FINTIM = 1.5, PRDEL = 0.05, OUTDEL = 0.05
PRINT X, XDOT, X2DOT
PRTPLT X
LABEL DISPLACEMENT VERSUS TIME
END
STOP
```

Figure 4-4.   CSMP III coding for the automobile suspension problem.

```
TIME           X
.0             .0
5.0000E-02     .20942     I---------+       I         I         I         I         I          I
.10000         .65543     I---------I---------I---------I--+      I         I         I          I
.15000         1.0819     I---------I---------I---------I---------I---------I---+      I          I
.20000         1.3290     I---------I---------I---------I---------I---------I---------I------+    I
.25000         1.3620     I---------I---------I---------I---------I---------I---------I--------+  I
.30000         1.2422     I---------I---------I---------I---------I---------I---------I-+         I
.35000         1.0691     I---------I---------I---------I---------I---------I---------I-+         I
.40000         .92886     I---------I---------I---------I---------I---------I----+     I          I
.45000         .86489     I---------I---------I---------I---------I---------I-+        I          I
.50000         .87583     I---------I---------I---------I---------I---------I---+      I          I
.55000         .93152     I---------I---------I---------I---------I---------I------+   I          I
.60000         .99472     I---------I---------I---------I---------I---------I--------+ I          I
.65000         1.0381     I---------I---------I---------I---------I---------I-+        I          I
.70000         1.0517     I---------I---------I---------I---------I---------I--+       I          I
.75000         1.0403     I---------I---------I---------I---------I---------I-+        I          I
.80000         1.0171     I---------I---------I---------I---------I---------I+         I          I
.85000         .99527     I---------I---------I---------I---------I---------+          I          I
.90000         .98281     I---------I---------I---------I---------I-------+I           I          I
.95000         .98138     I---------I---------I---------I---------I-------+I           I          I
1.0000         .98772     I---------I---------I---------I---------I-------+I           I          I
1.0500         .99667     I---------I---------I---------I---------I---------+          I          I
1.1000         1.0038     I---------I---------I---------I---------I---------+          I          I
1.1500         1.0070     I---------I---------I---------I---------I---------+          I          I
1.2000         1.0064     I---------I---------I---------I---------I---------+          I          I
1.2500         1.0034     I---------I---------I---------I---------I---------+          I          I
1.3000         1.0002     I---------I---------I---------I---------I---------+          I          I
1.3500         .99799     I---------I---------I---------I---------I---------+          I          I
1.4000         .99733     I---------I---------I---------I---------I--------+           I          I
1.4500         .99795     I---------I---------I---------I---------I--------+           I          I
1.5000         .99915     I---------I---------I---------I---------I--------+           I          I
```

Figure 4-5.   Print-plot output for the automobile suspension problem.

$K = 400$, as specified in the CONST statement. A PARAM statement has been used to produce runs with different values of D, to give different values to the damping ratio. The damping ratio values are 0.1, 0.3, 0.7, 1.0, and 2.0. The program integrates with an interval of 0.005, and runs for a time of 1.5.

The output is for tables of values of $\ddot{x}$, $\dot{x}$, and x, and a print-plot for x. All output is at intervals of 0.05. Figure 4-5 shows the print-plot output for the case of $D = 16.968$ ($\zeta = 0.3$). The curves of Fig. 1-9 were, in fact, plotted from the printed tables of this program.

4-9

Hybrid Simulation

For most studies the model is clearly either of a continuous or discrete nature, and that is the determining factor in deciding whether to use an analog or digital computer for system simulation. However, there are times when an analog and digital computer are combined to provide a *hybrid simulation*, (2). The form taken by hybrid simulation depends upon the application. One computer may be simulating the system being studied, while the other is providing a simulation of the environment in which the system is to operate. It is also possible that the system being simulated is an interconnection of continuous and discrete subsystems, which can best be modeled by an analog and digital computer being linked together.

The introduction of hybrid simulation required certain technological developments for its exploitation. High-speed converters are needed to transform signals from one form of representation to the other, (13). As a practical matter, the availability of mini-computers has made hybrid simulation easier, by lowering costs and allowing computers to be dedicated to an application. The term "hybrid simulation" is generally reserved for the case in which functionally distinct analog and digital computers are linked together for the purpose of simulation. It should not be confused with the use of digital elements added to the operational amplifiers of an analog computer, as discussed in Sec. 4-5, when describing hybrid computers.

4-10

Feedback Systems

A significant factor in the performance of many systems is that a coupling occurs between the input and output of the system. The term *feedback* is used to describe the phenomenum. A home heating system controlled by a thermostat is a simple example of a feedback system. The system has a furnace whose purpose is to heat a room, and the output of the system can be measured as room temperature. Depending upon whether the temperature is below or above the thermostat setting, the furnace will be turned on or off, so that information is being fed back from the output to the input. In this case, there are only two states, either the furnace is on or off.

An example of a feedback system in which there is continuous control is the aircraft system discussed in Sec. 1-1 and illustrated in Fig. 1-1. Here, the input is a desired aircraft heading and the output is the actual heading. The gyroscope of the autopilot is able to detect the difference between the two headings. A feedback is established by using the difference to operate the control surfaces, since change of heading will then affect the signal being used to control the heading. The difference between the desired heading θ_i and actual heading θ_o is called the error signal, since it is a measure of the extent to which the system deviates from the desired condition. It is denoted by ϵ.

Suppose the control surface angle is made directly proportional to the error signal. The force changing the heading is then proportional to the error and, consequently, it diminishes as the aircraft aproaches the correct heading. Consider what happens when the aircraft is suddenly asked to change to a new heading θ_i. The subsequent changes are illustrated in Fig. 4-6. The upper curves represent the control surface angle and the lower curves represent the aircraft heading.

Consider first the solid curves of Fig. 4-6. Upon receipt of a signal to change direction, the error signal, and therefore the control surface angle, suddenly takes a non-zero value. The aircraft heading, shown in the lower curves of Fig. 4-6, responds to the control surface by moving toward the new heading but, because of inertia, it takes time to respond. As the aircraft turns, the control surface angle decreases so that less force is applied as the aircraft approaches the required heading. Ultimately, when the aircraft reaches the required heading, the control surface angle will be zero but the inertia of the aircraft can carry it beyond the desired heading. As a result, the control surface turns in the opposite direction in order to bring the aircraft back from its overshoot. The correction from the overshoot produces an undershoot and the motion follows a series of oscillations of decreasing amplitude as illustrated in Fig. 4-6.

Suppose the system is changed so that for the same size error signal there is a larger control surface angle. The dotted curves of Fig. 4-6 represent this case. For the same conditions, the restoring force is larger and the aircraft responds more rapidly but, correspondingly, the initial overshoot will be larger. The aircraft will oscillate more widely and rapidly, as is illustrated in Fig. 4-6. The feedback loop in the second case is said to have greater *amplification*, since the same error produces a larger correction force. Under certain conditions, it is possible for the amplification to be so great that the initial overshoot exceeds the initial error; in which case, the undershoot from the correction becomes even larger and the system becomes unstable because of ever-increasing oscillations.

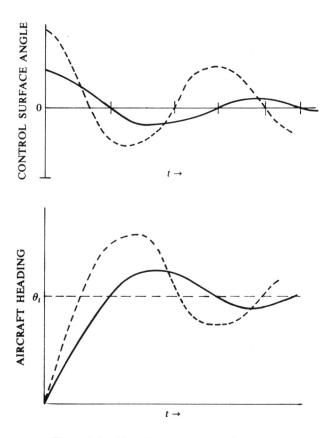

Figure 4-6. Aircraft response to autopilot system.

The feedback in the autopilot is said to be *negative feedback*. The more the system output deviates from the desired value the stronger is the force to drive it back.

Generally, negative feedback is a stabilizing influence, athough, as we have just seen, it can become strong enough to be a destabilizing influence. There are occasions when a system experiences *positive feedback*, in which case the force tends to increase the deviation. Any prolonged positive feedback will make a system unstable.

Many simulation studies of continuous systems are concerned primarily with the study of *servomechanisms*, which is the general name given to devices that rely upon feedback for their operation, (25). The field of *control theory* provides the theoretical background with which to design such systems, but simulation is extensively used to carry out detailed studies, (4). As an example, we will simulate the aircraft under the control of the autopilot.

4-11

Simulation of an Autopilot

In order to simulate the action of the autopilot, we first construct a mathematical model of the aircraft system. The error signal, ϵ, has been defined as the difference between the desired heading, or input, θ_i, and the actual heading, or output, θ_o. We therefore have the following identity:

$$\epsilon = \theta_i - \theta_o \tag{4-3}$$

We assume the rudder is turned to an angle proportional to the error signal, so that the force changing the aircraft heading is proportional to the error signal.

Instead of moving the aircraft sideways, the force applies a torque which will turn the aircraft. The strength of the torque, or turning force, depends on how far back the rudder is placed. However, just as the automobile movement was resisted by a shock absorber, the turning of the aircraft produces a resisting, viscous drag, approximately proportional to the angular velocity of the aircraft. The torque acting on the aircraft can, therefore, be represented by the following equation:

$$\text{Torque} = K\epsilon - D\dot{\theta}_o \tag{4-4}$$

where K and D are constants. The first term on the right-hand side is the torque produced by the rudder, and the second is the viscous drag.

When discussing differential equations in Sec. 4-2, it was stated that the fundamental law of mechanics is that the acceleration of a body is proportional to the applied force. That was related to linear motion. The same law, however, applies to a turning motion: the *angular* acceleration of a body is proportional to the applied torque. Further, the coefficient of proportionality is the inertia of the body, denoted by I. Since the angular acceleration of the aircraft is the second derivative of its heading, the equation of motion is

$$I\ddot{\theta}_o = \text{torque} \tag{4-5}$$

Substituting from Eqs. (4-3) and (4-4), and transposing terms, the resultant equation is

$$I\ddot{\theta}_o + D\dot{\theta}_o + K\theta_o = K\theta_i \tag{4-6}$$

If we divide both sides of the equation by I, and make the following substitutions

$$2\zeta\omega = \frac{D}{I}, \qquad \omega^2 = \frac{K}{I},$$

the equation of motion relating output to input then takes the following form:

$$\ddot{\theta}_o + 2\zeta\omega\dot{\theta}_o + \omega^2\theta_o = \omega^2\theta_i \qquad (4\text{-}7)$$

This is a second-order differential equation. Referring back to Sec. 1-10, it will be seen that it is in exactly the same form as Eq. (1-1), which describes the relationship between output and input for the automobile wheel suspension system.

Suppose the aircraft is initially flying a steady course which, by definition, we take to be the zero heading. If it is asked to change to a new heading at time zero, this corresponds to a unit step change of input. Because of the correspondence between the equations of motion we have just noted, the results will be the same as shown in Fig. 1-9, which showed the response of the suspension system to a unit step function. The results show that the aircraft will settle to the new heading, and the response will be oscillatory if ζ is less than 1.

To simulate how the autopilot can be designed to modify the aircraft response, it is more convenient to leave the model in the form of the three individual equations: Eqs. (4-3), (4-4), and (4-5). Using the variables ERROR and TORQUE to represent the error signal and the applied torque, the equations are

$$\text{ERROR} = \theta_i - \theta_o$$
$$\text{TORQUE} = K \star \text{ERROR} - D \star \dot{\theta}_o$$
$$I\ddot{\theta}_o = \text{TORQUE}$$

If we also use the variables HEAD, ANGVEL, and ANGACC to represent the aircraft heading and its first two derivatives, respectively, together with INPUT for the desired heading, the equations can be written

$$\text{ERROR} = \text{INPUT} - \text{HEAD}$$
$$\text{TORQUE} = K \star \text{ERROR} - D \star \text{ANGVEL}$$
$$I \star \text{ANGACC} = \text{TORQUE}$$

A CSMP III program for the system is shown in Fig. 4-7. Instead of using a step function change of heading, which simply repeats the automobile wheel case, the aircraft is being asked to turn in a circle. The desired heading is then continually increasing at a uniform rate so that

$$\text{INPUT} = A \star \text{TIME}$$

where A is a constant and TIME is a CSMP variable representing the time, t.[2] The constants K, I, and A have been set to values of 400, 2.00, and 0.0175, respectively. The constant D has been programmed to take different values on five separate runs, so that the damping ratio, ζ, will have the values 0.1, 0.3, 0.7, 1.0, and 2.0.

---

[2]With t expressed in seconds, the equation of motion gives θ_o in radians. The value chosen for A in the simulation makes the input a request for a turn of one degree a second.

```
TITLE AIRCRAFT WITH RATE CONTROL
*
PARAM D = (5.656, 16.968, 39.592, 56.56, 113.12)
 INPUT = A*TIME
 ERROR = INPUT - HEAD
 TORQUE = K*ERROR - D*ANGVEL
 ANGACC = TORQUE/I
 HEAD = INTGRL(0.0,ANGVEL)
 ANGVEL = INTGRL(0.0,ANGACC)
*
CONST I = 2.0, K = 400.0, A = 0.0175
TIMER DELT = 0.005, FINTIM = 1.5, PRDEL = 0.05
PRINT HEAD
LABEL HEADING VERSUS TIME
END
```

Figure 4-7. CSMP III program for aircraft with rate control.

The results of the runs are plotted in Fig. 4-8. They are given in non-dimensional form by plotting $\omega\theta_o/A$ against ωt. The straight line through the origin is the desired heading, which would be followed by the aircraft if it responded perfectly. However, it can be seen that the aircraft lags behind the desired heading and never catches up so long as the turn request remains in effect. The size of the lag increases as the damping ratio increases.

The error signal appears as an electrical signal in the autopilot. It can easily be

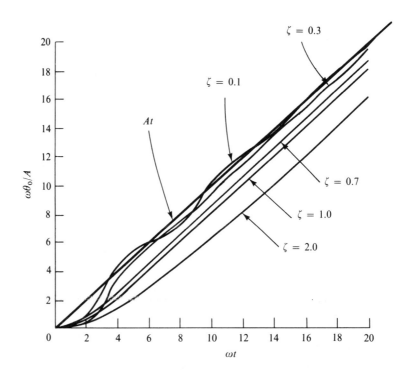

Figure 4-8. Response of aircraft flying in a circle with rate control.

modified by electrical circuits. The motion of the aircraft can also be measured by instruments giving electrical signals that can be added to the autopilot output. The modifications affect the aircraft response by changing the applied force.

Studies in control theory are concerned with understanding how system behavior can be modified in this way. As an example, suppose it is important to eliminate the lag that builds up as the aircraft turns. One way this can be done is to add to the error signal a component that is proportional to the integral of the error signal. The applied force is then as follows:

$$F(t) = K\epsilon + K_1 \int \epsilon \, dt$$

With this modification, control theory shows that for the system to remain stable, we must have

$$K_1 < \frac{KD}{I}$$

Figure 4-9 shows the response of the aircraft, under the same conditions as were used before, when this modification is included. In each case, the value of K_1 was

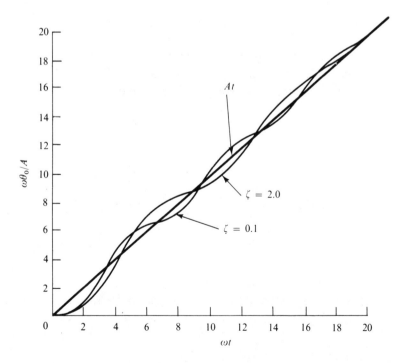

Figure 4-9. Response of aircraft flying in a circle with integral control.

chosen so that

$$\frac{K_1}{K} = \frac{0.2D}{I}$$

The results of Fig. 4-9 show that the second modification, which is called *integral control*, will eliminate the lag when the aircraft is performing a steady turn.

4-12
Interactive Systems

While all system simulation, as we have defined it, is concerned with the response of a system over time, the observation of the response over time is particularly important for continuous system simulation. Whereas the output of a discrete system simulation is likely to be a specific measure, such as a mean waiting time, the result of a continuous system simulation is likely to be judged by such factors as whether the system is stable or whether its output oscillates. Visual inspection of the results as they develop is therefore very valuable. Most CSSLs provide the ability to display the outputs on a screen, or a plotter. In particular, CSMP III provides this capability, (15).

Given this capability, a user can judge whether the output is developing satisfactorily, and, if not, interrupt a simulation run to try for better conditions, (9) and (10). Most graphic terminals serve as both input and output devices by providing a keyboard for entering data, and, perhaps, a lightpen which can be used to identify a particular output to the computer, (24), and (27). It is a simple matter, therefore, to instruct the computer to take another case.

4-13
Real-time Simulation

We illustrated in Sec. 4-2 the general process by which a mathematical model of an engineering system is constructed. It involves understanding physical or chemical laws, and implies a number of experiments or measurements to derive the coefficients of the model. This can be particularly time-consuming if the model is not being simplified by assuming linearity.

It must have occurred to the reader that this detailed, preliminary work could be avoided if an actual device can be used, rather than constructing a model. This approach is, in fact, followed when using *real-time simulation*. With this technique, actual devices, which are part of a system, are used in conjunction with either a digital or hybrid computer, providing a simulation of the parts of the system that do not exist or that cannot conveniently be used in an experiment. Real-time simulation will often involve interaction with a human being, thereby avoiding the need to design and validate a model of human behavior.

Along with the technological developments needed for hybrid simulation, real-time simulation requires computers that can operate in real-time; this means that they must be able to respond immediately to signals sent from the physical devices, and send out signals at specific points in time. These capabilities have been available for many years, and so real-time simulation is used in many areas. The aerospace industry in particular makes extensive use of real-time simulation, (16) and (18).

There are devices called *simulators* (12), whose main function is to provide human beings with a substitute for some environment or situation, for example, devices for training pilots by giving them the impression they are at the controls of an aircraft. A well known example was used to train astronauts. They were suspended in spring harnesses which reduced their apparent weight to that which would occur on the moon surface. Since the main purpose of such devices is to duplicate some sort of sensory stimulus, they cannot properly be classified as computers.

Exercises

Derive a CSMP III program for the following problems.

4-1   Modify the automobile suspension program of Sec. 4-8 so that the displacement of the mass cannot exceed 1.25.

4-2   Reprogram the automobile suspension problem with the assumption that the damping force of the shock absorber is equal to $5.656(\dot{x} - 0.05\ddot{x})$.

4-3   Solve the following equations:

$$3\dddot{x} + 15\ddot{x} + 50\dot{x} + 200x = 10$$
$$\dddot{x} = \ddot{x} = \dot{x} = x = 0 \qquad \text{at } t = 0$$

4-4   A torsional pendulum consists of a bar suspended from a spring that winds and unwinds as the bar oscillates up and down. If x is the vertical displacement and θ is the angle the spring turns, the following equations describe the motion:

$$a\ddot{x} + bx - c\theta = F(t)$$
$$e\ddot{\theta} + f\theta - gx = G(t)$$

Assume the system starts from rest with x and θ equal to zero, $F(t)$ remains zero for all time and $G(t)$ is 1 for $t \geq 0$. Solve the equations for the following values:

$$a = e = 1$$
$$b = f = 40$$
$$c = g = 4$$

4-5 In a chemical reaction one molecule of a substance X is produced for one molecule each of substances A and B. The initial concentrations of A and B are a and b, respectively. Let x be the concentration of X and assume that it is initially zero. The rate at which x increases is 0.1 times the product of the current concentrations of A and B. Assume a and b are initially 0.8 and 0.4, respectively, and simulate the production of X.

Bibliography

1 BARNA, ARPAD, AND DAN I. PORAT, *Integrated Circuits in Digital Electronics*, New York: John Wiley & Sons, Inc., 1973.

2 BEKEY, G. A., AND W. J. KARPLUS, *Hybrid Computation*, New York: John Wiley & Sons, Inc., 1968.

3 BRENNAN, R. D., AND M. Y. SILBERBERG, "Two Continuous System Modeling Programs," *IBM Syst. J.*, VI, no. 4 (1967), 242–266.

4 BREWER, JOHN W., *Control Systems: Analysis, Design, and Simulation*, Englewood Cliffs, N. J.: Prentice-Hall, Inc., 1974.

5 BUSH, V., "The Differential Analyzer: A New Machine for Solving Differential Equations," *J. Franklin Inst.*, CCXII, no. 4 (1931), 447–488.

6 CARDENAS, ALFONSO F., AND WALTER J. KARPLUS, "PDEL-A Language for Partial Differential Equations," *Commun. ACM*, XIII, no. 3 (1970), 184–191.

7 CHINH, NGUYEN T., AND MIGUEL A. MARIN, "Real Roots of Algebraic Equations by Hybrid Computer Simulation," *Simulation*, XXI, no. 6 (1973), 181–183.

8 CHU, YOAHAN, *Digital Simulation of Continuous Systems*, New York: McGraw-Hill Book Company, 1969.

9 COMBA, PAUL, "A Language for Three-Dimensional Geometry," *IBM Syst. J.*, VII, nos. 3 & 4 (1968), 292–307.

10 GAGLIANO, F. W., H. W. THOMBS, AND R. E. CORNISH, "A Conversational Display Capability," *IBM Syst. J.*, VII, nos. 3 & 4 (1968), 281–291.

11 GREBENE, ALAN B., *Analog Integrated Circuit Design*, New York: Van Nostrand Reinhold Company, 1972.

12 HALL, G. WARREN, "Recent Advances in In-Flight Simulator Technology," *Proc. Eighth Annual Simulation Symposium*, 1975, 11–29. Annual Simulation Symposium, P. O. Box 22573, Tampa, Fla.

13 HOESCHLE, JR., D. F., *Analog-to-Digital—Digital-to-Analog Conversion Techniques*, New York: John Wiley & Sons, 1968.

14 IBM Corp., CSMP III Program Reference Manual, Form no. SH 19–7001 (1972), White Plains, N. Y.

15 ————, CSMP III Graphic Feature, Program Reference Manual, Form no. SH 19–7003 (1972).

16 KORN, GRANINO A., "Project DARE-Differential Analog Replacement by On-Line Digital Simulation," *Proc. AFIPS Joint Computer Conf.*, vol. 35, pp. 247–254. Montvale, N. J.: AFIPS Press, 1969.

17 ————, AND THERESA M. KORN, *Electronic Analog and Hybrid Computers*, New York: McGraw-Hill Book Company, 1972.

18 LANGE, A. S., "Semi-Physical Simulation of Guided Missiles," *Comp. and Electr. Eng.*, I, no. 1 (1973), 119–142.

19 LINEBARGER, ROBERT N., AND ROBERT D. BRENNAN, "A Survey of Digital Simulation: Digital-Analog Simulation Programs," *Simulation*, III, no. 6 (1964), 22–36.

20 MADICH, P., J. PETRICH, AND N. PAREZANOVITCH, "The Use of a Repetitive Differential Analyzer for Finding Roots of Polynomial Equations," *IRE Trans. Electron. Comput.*, vol. EC-8, no. 2 (1959), 182–185.

21 MURRAY, FRANCIS J., *Mathematical Machines*, vol. 2 *Analog Devices*, New York: Columbia Press, 1961.

22 SPECKHART, FRANK H., AND WALTER L. GREEN, *A Guide to Using CSMP*, Englewood Cliffs, N. J.: Prentice-Hall, Inc., 1976.

23 STRAUSS, J. C., D. C. AUGUSTINE, B. B. JOHNSON, R. N. LINEBARGER, AND F. J. SANSON, "The SCi Continuous System Simulation Language, (CSSL)," *Simulation*, IX, no. 6 (1967), 281–303.

24 TEICHOLZ, ERIC, "Interactive Graphics Comes of Age," *Datamation*, XXI, no. 12 (1975), 50–53.

25 THALER, GEORGE J., *Design of Feedback Systems*, New York: Halsted Press, 1973.

26 VOLZ, RICHARD A., "Examples of Function Optimization Via Hybrid Computation," *Simulation* XXI, no. 2 (1973), 43–48.

27 WALKER, B. S., *Interactive Computer Graphics*, New York: Crane Russack, Inc., 1976.

5

SYSTEM DYNAMICS

5-1

Historical Background

Control theory is concerned with understanding how the response of a system can be changed by modifying the signals occurring in the system. The autopilot study of Sec. 4-11 gave a simple example. As was seen, maintaining stability and controlling oscillations are prime concerns.

There are many examples outside the field of engineering in which some form of instability or oscillation can occur. Business cycles cause fluctuations in the general level of economic activity. In individual industries there are similar cycles where prices and supplies fluctuate or temporarily appear to go out of control. There are also many examples in the field of biology of oscillations, or explosions, in populations. It seems natural to speculate on whether these phenomena can be controlled. Clearly, the precision obtained in engineering cannot be duplicated in economic or social systems. Nevertheless, insight gained from control theory can be used to analyze the systems and, perhaps, suggest changes that will improve performance.

Initial studies of this nature dealt with industrial systems, and were called *Industrial Dynamics* studies, (8). The dramatic changes in urban centers that began to occur in the 1950s led to *Urban Dynamics* studies, (9). More recently, there has been great concern about many social and environmental problems, which, it can be seen, are world-wide in their scope. Study of these problems has been the subject of *World Dynamics*, (10). Since there are no major differences in the techniques used in these different areas, the general name *System Dynamics* has come to be used.

The principal concern of a System Dynamics study is to understand the forces operating in a system in order to determine their influence on the stability or growth of the system. The output of the study will, it is hoped, suggest some reorganization, or change in policy, that can solve an existing problem or guide developments away from potentially dangerous directions. It is not usually expected that a System Dynamics study will produce specific numbers for redesigning a system, as occurs with engineering systems. Correspondingly, many of the coefficients in the models of System Dynamics studies consist of estimates or best guesses, particularly since the models must sometimes reduce such qualitative factors as personal preferences, or social tension, to quantitive form. Nevertheless, the lack of precision that may have to be tolerated does not destroy the value of the study. The model can establish the relative effectiveness of different policies under the same assumptions, or mark out ranges of values that can be expected to produce a given type of output.

5-2

Exponential Growth Models

Growth implies a rate of change. Consequently, mathematical models describing growth involve differential equations. Consider, for example, the growth of a capital fund that is earning compound interest. If the growth rate coefficient is k (i.e., $100k\%$ interest rate), then the rate at which the fund grows is k times the current size of the fund. Expressed mathematically, where x is the current size of the fund,

$$\dot{x} = kx$$
$$x = x_0 \quad \text{at } t = 0$$

(5-1)

This is a first-order differential equation whose solution is the exponential function. The solution, in terms of the mathematical constant e, which has the approximate value 2.72, is

$$x = x_0 e^{kt}$$

Figure 5-1 plots x for various values of k and an initial value of 1. It can be seen that the fund grows indefinitely, whatever (positive) value of k is used, and it grows faster with greater values of k. Looking at the curve for $k = 0.2$, and picking the point $x = 2$, the corresponding slope at the point A has a certain value. Later,

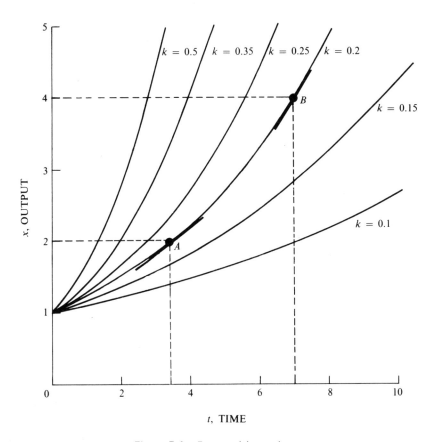

Figure 5-1. Exponential growth curves.

at the point B, where x has become twice its value at A, the slope has become twice as great as at A. Since the slope is measured as the first-order differential coefficient, this fact is simply the reflection of the law defining exponential growth: the growth rate is directly proportional to the current level.

Another way of describing the exponential function is to say that the logarithm of the variable increases *linearly* with time. To test whether any particular set of data represents exponential growth, the logarithms of the data should be plotted against time. If the data then appear to fall on a straight line, the growth is exponential, and the slope of the straight line will be greater for larger growth rate coefficients.

Alternatively, the data can be plotted on semi-logarithmic graph paper where the horizontal lines are placed at logarithmic intervals. Plotting data on such paper is equivalent to taking the logarithm of the data and then plotting on normal linear graph paper. For example, Fig. 5-2 shows the gross national product figures

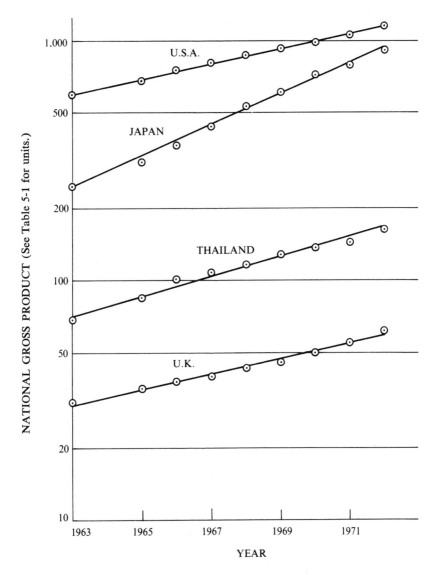

Figure 5-2. Growth of gross national products.

for several countries plotted against year on semi-logarithmic paper, (21).[1] The points fall reasonably well on straight lines, indicating exponential growth rates.

The growth rate coefficient can be estimated by picking two points of the straight line that best fits the data, and taking the (natural) logarithm of the ratio of the

[1]The lines of the figure have been drawn freehand.

values. If the points are x_1 at t_1 and x_2 at t_2 $(t_2 > t_1)$, the result is

$$\ln \frac{x_2}{x_1} = (t_2 - t_1)k$$

from which it is possible to derive k. In terms of the more familiar logarithms to the base 10, the corresponding result is

$$\log \frac{x_2}{x_1} = 0.434(t_2 - t_1)k$$

For example, looking at the figures for the U.K., in Table 5-1, the values for 1965 and 1971 are 35.3 and 55.6, respectively. The ratio is 1.575, and, since the logarithm to the base 10 of this number is 0.1973, the growth rate coefficient is 0.085, or $8\frac{1}{2}\%$.[2]

Sometimes the coefficient k is expressed in the form $1/T$, so that

$$k = \frac{1}{T}$$

The solution for the exponential growth model then takes the form

$$x = x_0 e^{t/T}$$

TABLE 5-1 **National Gross Domestic Products**

Year	UK	USA	Thailand	Japan
1963	30.2	596	68.1	245
1965	35.3	692	84.3	321
1966	37.7	759	101.4	369
1967	39.8	804	108.3	437
1968	42.9	863	116.7	518
1969	45.7	929	128.6	605
1970	49.9	983	135.9	712
1971	55.6	1060	145.3	794
1972	61.1	1159	160.2	907

Figures are in units of milliards (10^9) of the countries' monetary units, except for Japan's, which are in hundreds of milliards yen.

The constant T is said to be a *time constant* since it provides a measure of how rapidly the variable x grows. For example, when t equals T, the variable is exactly

[2]The data are not adjusted to allow for inflation.

e times its initial value x_0. If T is small, say 2, x reaches this level after two time units. If T is large, say 20, x only reaches that level after 20 time units. The level of e times the initial value has only been chosen as a convenient example. Whatever level is chosen as a basis of comparison, the ratio of the time constants of two different exponential growth models will give the relative times that the two systems will take to reach that level.

The inverse relationship between k, the growth rate coefficient, and T, the time constant, means that a large coefficient is associated with a small time constant and, therefore, a more rapid rate of increasing. For example, a growth rate of 0.05, or 5%, will double the size of a population in a little under 14 years, while a growth rate of 10% will do the same in just under 7 years. Doubling the growth rate again, to 20%, will reduce the time to $3\frac{1}{2}$ years. We can say the values 14, 7, and $3\frac{1}{2}$ are the time constants of the three cases.

5-3

Exponential Decay Models

Another model, closely related to the exponential growth model, is a model in which a variable decays from some initial value, x_0, at a rate proportional to the current value. The model can, in fact, be interpreted as a negative growth model. The equation for the model is

$$\dot{x} = -kx \qquad\qquad (5\text{-}2)$$
$$x = x_0 \quad \text{at } t = 0$$

The solution is

$$x = x_0 e^{-kt}$$

The response is shown in Fig. 5-3 for various values of k. As with the exponential growth model, the constant k is sometimes expressed in the form $1/T$. The characteristic of the model is that the level, x, is *divided* by a constant factor for a given interval of time. In the interval of T time units, the level is divided by e. Since e is approximately 2.72, the level is reduced by a factor of 0.37. Each successive interval of T reduces the level by the same factor. Again, there is nothing significant about the value e, but comparing values of T for different models measures the relative times they will take to decay by a given fraction.

An example of a decay model is an aging population, which is being diminished by deaths or breakdowns. Another example is the manner in which radioactive material decays.

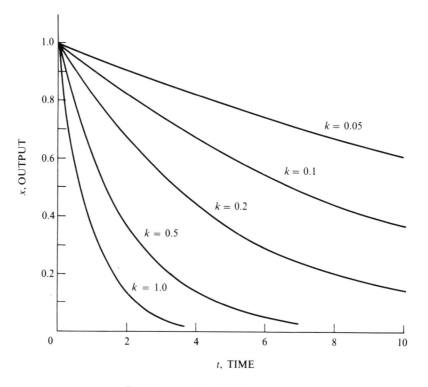

Figure 5-3. Exponential decay curves.

5-4
Modified Exponential Growth Models

A market model attempts to describe how many items of a product will be sold. In practice, there is a limit to the market, if we ignore replacement sales. The exponential growth model, therefore, is not a satisfactory market growth model because it shows an unlimited growth.

The exponential growth model assumes the rate of growth is proportional to the existing level. A more realistic assumption for a market model is that the growth rate is proportional to the number of people who have not yet bought the product. Suppose the market is limited to some maximum value, X: the number of people who might buy the product, in our example. Let x be the number of people who *have* bought the product. Then the number who have not is $X - x$, and, if we use k as the coefficient of proportionality, the equation for the model is

$$\dot{x} = k(X - x)$$
$$x = 0 \quad \text{at } t = 0$$

(5-3)

and the solution is

$$x = X(1 - e^{-kt})$$

Figure 5-4 plots the solution for a number of values of k. This type of curve is sometimes referred to as a *modified exponential* curve. As can be seen, the maximum slope occurs at the origin and the slope steadily decreases as time increases. As a result, the curve approaches the limit more slowly the closer it gets, and never actually reaches the limit. In marketing terms, the sales rate drops as the market

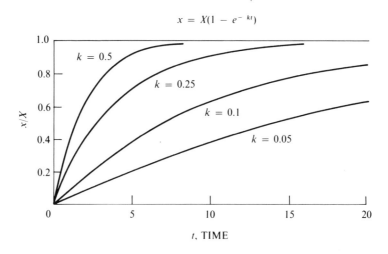

$$x = X(1 - e^{-kt})$$

Figure 5-4. Modified exponential curves.

penetration increases. The constant k plays the same role as the growth rate coefficient in the exponential growth model. As k increases, the sales grow more rapidly. As with the exponential model, the constant is sometimes expressed in the form $1/T$, in which case, it can be interpreted as a time constant.

5-5
Logistic Curves

An unrealistic feature of the modified exponential curve, considered as a market model, is that it shows the maximum slope at the beginning. In practice, the sales rate, which is represented by the slope, begins at a low value and initially increases, as occurs in the exponential growth model. Eventually the slope begins to decline, making the curve of market growth look more like the *modified* exponential curve. The result is an S-shaped curve resembling the one shown in Fig. 5-5. Initially the curve turns upwards, looking like the exponential growth model curve. Eventually

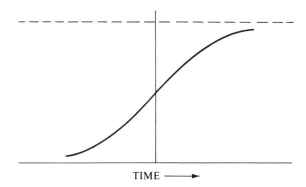

TIME ———→

Figure 5-5. S-shaped growth curve.

the curve reaches a point of inflection, where it turns from an increasing to a decreasing slope. From there, the curve resembles a modified exponential curve.

An explanation of this shape, in terms of the behavior of people buying a product, can be given. Initially, when the product is new and unknown to most people, knowledge about the product will tend to spread by word-of-mouth, or by potential customers actually seeing one of the products that has been bought. Under those circumstances, the sales rate will tend to be proportional to the number sold, which is the condition governing the exponential growth curve. As more products are sold, and the market begins to become saturated, the conditions of the *modified* exponential model take over.

The *logistic function* is, in effect, a combination of the exponential and modified exponential functions that describes this process mathematically. The differential equation defining the logistic function is

$$\dot{x} = kx(X - x) \qquad\qquad (5\text{-}4)$$

Initially, when x is very much smaller than X, which is the market limit, the value of the term $X - x$ remains essentially constant at the value X. The equation for the logistic function, therefore, is approximately

$$\dot{x} = kXx$$

which is the equation for the exponential growth model with a constant of kX. Much later, when the market is almost saturated, the value of x will be close to the value of X, so that it changes very little with time. The equation for the logistic function then takes the approximate form,

$$\dot{x} = kX(X - x) \qquad\qquad (5\text{-}5)$$

which is the differential equation for the modified exponential function with a constant of kX.

The true differential equation for the logistic function is nonlinear, because it contains the term x^2. Its solution is relatively complicated, and it will not be given here. The reader is referred to reference (7), which gives the full solution and discusses the methods of fitting the function to actual data.

The logistic function has been described as a market model. However, it is also applicable to many other types of systems. Population growth, for example, can follow a logistic curve, (13). The initial rapid growth corresponds to the situation in which there are ample resources to support life, and the eventual leveling of growth corresponds to the resources becoming fully utilized.

The model can also represent the spread of a disease. The initial rapid growth then corresponds to the disease being transmitted by contact between infected and uninfected subjects. The leveling of growth occurs as the number of subjects as yet uninfected drops.

5-6

Generalization of Growth Models

We have now seen three basic growth models: the exponental, modified exponential, and logistic. Many other mathematical forms could be used as a basis for a model. In particular, each of the basic models depends upon certain coefficients, which have been assumed to be constant. The models can be generalized by removing this assumption. This is especially true of the models that assume a market limit. In practice, such a limit may exist but it is not likely to remain constant. Instead, it is likely to expand with population growth, or fluctuate with economic conditions, or respond to both these factors.

An example is a model that shows the growth in the number of residential telephones in the United States, (5). Here, an underlying logistic model was assumed, but the residual market was assumed to depend upon three economic factors: the number of households in the U.S., denoted by H_t; disposable personal income per capita, P_t; and local service revenue per telephone, C_t. The model takes the form (in the present notation)

$$\dot{x} = kx(X - x) \qquad (5\text{-}6)$$

$$(X - x) = aH_t^{\alpha_1} P_t^{\alpha_2} C_t^{\alpha_3} \qquad (5\text{-}7)$$

After finding the best values of the constants to fit the data to the model, Fig. 5-6[3]

[3] Adapted from Ref. (5) with permission of The Bell Journal of Economics and Management Science. © 1971 by the American Telephone and Telegraph Company.

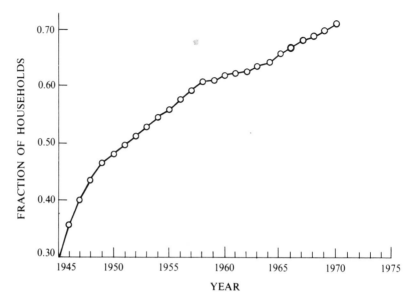

Figure 5-6. Bell system residential main telephones per U.S. (continental)
household.

shows how the model, shown as a solid line, fits the data, shown as circles. (Note
that the figure is confined to the upper portion of a logistic curve.)

5-7
System Dynamics Diagrams

A System Dynamics view of a system concentrates on the rates at which various
quantities change and expresses the rates as continuous variables. The flow of
orders from customers will be described as a certain rate of orders and the delivery
of orders will similarly be represented by a rate. The difference between the two
rates will be integrated to give the level of unfilled orders at any time.

The basic structure of a System Dynamics model is illustrated in Fig. 5-7. It
consists of a number of reservoirs, or levels, interconnected by flow paths. The
rates of flow are controlled by decision functions that depend upon conditions in
the system. The *levels* represent the accumulation of various entities in the system,
such as inventories of goods, unfilled orders, number of employees, etc. The current
value of a level at any time represents the accumulated difference between the input
and the output flow for that level. *Rates* are defined to represent the instantaneous
flow to or from a level. *Decision functions* or, as they are also called, *rate equations*
determine how the flow rates depend upon the levels.

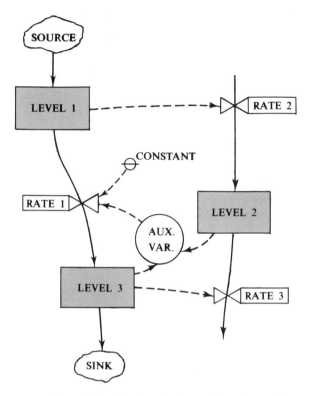

Figure 5-7. Structure of a System Dynamics model.

A set of symbols has been established for indicating the various factors involved in a System Dynamics model and some of these are illustrated in Fig. 5-7. Levels are represented by boxes; and rates are indicated by a symbol which, as a matter of interest, is adapted from the symbol used in diagrams of hydraulic systems to represent a valve. It is sometimes necessary to indicate a source or sink of entities, as is shown in Fig. 5-7. For completeness, the need for a constant in the model is indicated with a special symbol. Sometimes an auxiliary variable needs to be defined, usually by combining other variables mathematically by some relationship other than the integration implied by a rate equation and a level. Auxiliary variables are depicted by circles. Solid lines in the diagram indicate the flow of tangible objects. Dotted lines indicate causal relationships.

Levels would appear to be dimensionless quantities since they represent a count. However, the term level is also applied to some quantities that do have dimensions. For example, the rate of reordering goods may depend upon the average number of orders over some period of time, which is a quantity that has the dimensions of number over time. Nevertheless, the average number of orders per week would be regarded as a level in a System Dynamics model. The simplest test for determining which quantities are to be regarded as levels is to imagine the system brought to rest. Any quantity that is a rate is then automatically zero, while any quantity that maintains a magnitude should be regarded as a level.

5-8

Simple System Dynamics Diagrams

The simple exponential growth model is concerned with a single level representing, say, a population. The model would be represented by the System Dynamics diagram (a), shown on the left of Fig. 5-8. The level, x, representing the population, is being increased by a single rate factor, representing the excess of the birth rate over the death rate. The rate equation is Eq. (5-1), that is $\dot{x} = kx$. The dependence

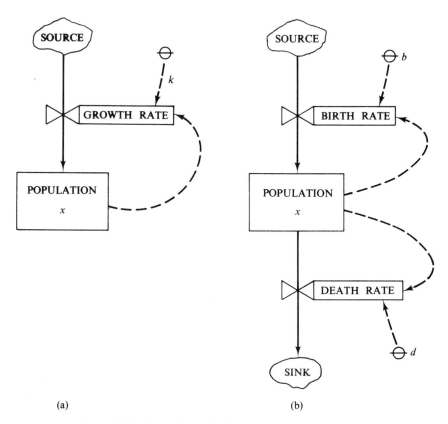

(a)   (b)

Figure 5-8. System Dynamics diagrams of population growth.

of the rate on the level, x, is indicated by the dashed line of Fig. 5-8 (a). The coefficient k is shown as a constant.

An alternative way of modeling the population is to represent the births and deaths separately, as is done in diagram (b) of Fig. 5-8. The level is now affected by two rate variables, one for births to increase the level, and one for deaths to decrease the level. If we assume that both rates depend upon the current level of the population, the rate equation for the births is the exponential growth model of Eq. (5-1) with the birth rate coefficient, b, and the rate equation for the deaths is the decay model of Eq. (5-2) with the coefficient set to the death rate, d.

If we consider the market models that were discussed before, we could use the two diagrams shown in Fig. 5-9. Diagram (a), on the left, represents the modified

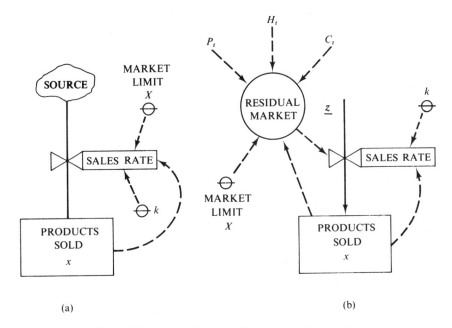

Figure 5-9. System Dynamics diagrams of market models.

exponential model. A constant X, the market limit, has been introduced, and the rate is shown as depending upon the constant k and the current level of the market. The rate equation is Eq. (5-3).

The logistic model could be represented in exactly the same way, with the rate equation being given by Eq. (5-4). However, if the model is generalized in the manner discussed in Sec. 5-6, where the remaining market depends upon three economic variables, P_t, H_t, and C_t, an auxiliary variable is needed to represent the residual market. This is indicated in diagram (b) of Fig. 5-9, where it is implied that the three economic factors are derived from levels that are not shown. If the auxiliary variable is denoted by z, it is defined by the right-hand side of Eq. (5-7), and the rate, $\dot{x}$, affecting the level is kzx.

5-9
Multi-Segment Models

In Sec. 3-6 we looked at a model of house air-conditioner sales to illustrate the general method of carrying out the numerical computations involved in a continuous system simulation. The model was as follows:

H Number of households
y Number of houses sold
x Number of air conditioners installed

$$\dot{y} = k_1(H - y) \tag{5-8}$$
$$\dot{x} = k_2(y - x) \tag{5-9}$$

The first equation is a modified exponential growth model showing how the number of houses increases, with a fixed limit H. The second equation is also a modified exponential growth model for the number of air conditioners, but it has a variable limit, y. Figure 5-10 shows a System Dynamics diagram of the model.

Suppose we wish to extend the model by considering the breakdown of air conditioners. Then, a decay model, where it is assumed that the decay rate depends

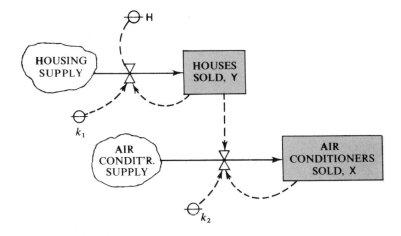

Figure 5-10. System Dynamics model of air-conditioner sales.

upon the current level, needs to be added to diminish the level of x. Suppose also, we no longer take the housing limit to be fixed. Instead, assume an exponential growth model for the number of households. The expanded system can now be represented by the diagram of Fig. 5-11. A new equation needs to be given for x, and an equation must be added for H, as follows:

$$\dot{x} = k_2(y - x) - k_3 x \qquad (5\text{-}10)$$

$$\dot{H} = k_4 H \qquad (5\text{-}11)$$

The complete model consists of Eqs. (5-8), (5-10), and (5-11).

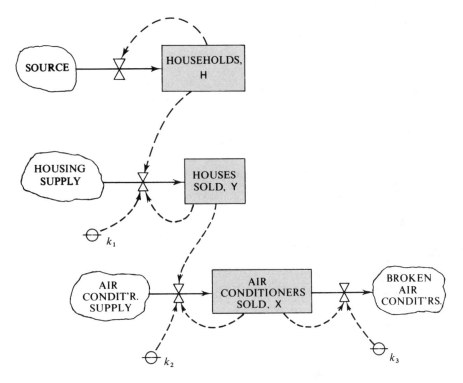

Figure 5-11. Extended air-conditioner sales model.

In Chap. 2 we saw in the corporate model how a complex model is constructed from individual segments. The present simple model could be developed in the same way. The individual levels are representative of the national economy, as it affects potential markets, consumer behavior, and product life. The model could be further developed by making more detailed models of these segments, or by replicating segments of the model to represent, simultaneously, different geographic regions, various products, and a range of social groups. System Dynamics models of this nature are well suited to studying the interactions that are typical of socio-economic systems, (12).

5-10

Representation of Time Delays

It was pointed out in Sec. 5-2 that the coefficients involved in growth models can be interpreted as time constants. In the air-conditioner model, for example, the constants k_1, k_2, and k_3 can be expressed as $1/T_1$, $1/T_2$, and $1/T_3$. The model can then be described as representing the fact that there is a time lag of T_1 time units before changes in the economy affect buying trends, that there is a lag of T_2 units before new houses get air conditioners, or that the average life of an air conditioner is T_3 units.

When there is a need to represent a time delay, the simplest method is to introduce an appropriate time constant. For example, suppose there is a level x of outstanding orders, and it is known that it takes an average of D time units to fill an order. A System Dynamics representation would say that the outstanding-order level is being depleted with a time constant of D time units, and would include the following decay equation:

$$\dot{x} = \frac{-x}{D}$$

In general, to represent a delay, it is necessary to identify the level that is controlled by the delay and apply the rule

$$\text{Delay} = \frac{\text{Level}}{\text{Rate}}$$

This is sometimes called an *exponential delay*, because of the solution it produces. [See Eq. (5-2).] This is a very over-simplified way of representing time delays, but it is consistent with the character of a System Dynamics study that is trying to comprehend the general behavior of a system rather than aim for precision. If the model is programmed in a CSSL such as CSMP, there is no difficulty in introducing a finite delay of fixed size, because most CSSLs allow the use of finite delays. Even if the programming language only allows continuous equations, it is possible to approximate a finite delay, more closely by using a series of exponential delays. A pair of delays of size $T/2$ placed in series is closer to a finite delay than one delay of size T, and the fit becomes better as this process is continued, (17).[4]

5-11

Feedback in Socio-Economic Systems

Feedback occurs in socio-economic systems in much the same way it occurs in physical systems. The feedback is not usually caused by a physical signal, but occurs through the feedback of information. Some decision is made on the basis of information about prevailing conditions, and the action taken as a result of the decision changes the conditions upon which the information is based. The link between the decision and the information may go unnoticed. There could be a long chain of circumstances connecting the two. There has, for example, been much speculation that feedback causes business cycles in the economic growth of a country, and many efforts have been made to identify the underlying forces. On the other hand, feedback may deliberately be introduced in a socio-economic system to help control the system.

A feedback system that is analogous to the autopilot system, discussed previously in Sec. 4-11, is seen in the control of inventories. A retailer will usually maintain an inventory of goods, aiming to keep the level such that it strikes a reasonable balance between the cost of holding goods in inventory and the penalty of losing sales if the inventory should become empty.

The diagram of an autopilot system in Fig. 1-1, could also represent an inventory control system. The input θ_i that represented desired heading represents the desired inventory level. The output θ_o that represented actual heading now represents actual inventory level. The difference, or, as it was previously called, error signal, controls the orders placed to replenish the inventory. The delay, represented by the inertia of the aircraft, is the delay in filling the orders. In a manner analogous to the aircraft heading, the inventory level can oscillate and it is even possible that, if the retailer overreacts by ordering large quantities for small differences, the system will correspond to the autopilot system with too much amplification and will be unstable.

In practice, an inventory system does not usually have a negative error signal; that is to say, when the inventory exceeds the desired level, the orders do not go

---

[4]Readers familiar with electrical engineering will recognize that the exponential decay model is the equation for the simple $R - C$ filter that is used to delay a signal. It is usually called a delay circuit, and its time constant is defined as the product RC. Further, a coaxial cable, which, electrically, produces a finite delay, can be approximated by a large number of $R - C$ filters in series.

negative. If they did, it would correspond to saying that goods are sent back. It is more likely that there is a steady drain on the inventory which corresponds to normal sales. The retailer establishes a steady average order *rate* which balances the average rate of sales and maintains a constant inventory level. The changes represent changes with respect to the average order rate, so that a negative error signal corresponds to reducing the order rate below the normal amount.

A simple model of the inventory control system would need two levels and three rates defined as follows:

X Current inventory level
Y Outstanding level of orders placed with the supplier
U Rate of ordering from supplier
V Rate of delivery from supplier
S Rate of sales

In addition, two constants need to be defined:

I Planned inventory level
T_1 Average delivery time

A model of the system is shown in Fig. 5-12. It shows the level of outstanding orders being controlled by the rate at which orders are placed and the rate of delivery. Similarly, the level of the inventory is controlled by the rate of delivery and the rate of sales.

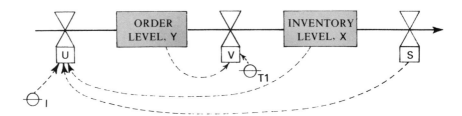

Figure 5-12. System Dynamics model of an inventory control system.

To derive a mathematical model of the system, equations can immediately be written for the two levels. These are

$$\dot{Y} = U - V$$
$$\dot{X} = V - S$$

Using the general formula for delays, the equation for the rate of delivery can be written

$$V = \frac{Y}{T_1}$$

The rate of ordering is a management decision. Clearly, it should balance the steady sales rate, but simply to equate order rate to sales rate is not satisfactory. When the sales rate increases, the inventory will become depleted because the delivery rate takes time to catch up with the new sales rate. Similarly, a drop in sales rate will result in the inventory level rising. The decision, therefore, is to add to the sales rate a term proportional to the difference between the desired level and the actual level. When the inventory is low, this increases the order rate and, when the level is high, it decreases the rate. The order rate is, therefore,

$$U = S + K(I - X)$$

where K is a constant that will be called the *ordering constant*.

A CSMP III program for the model is shown in Fig. 5-13. Assume that time is measured in days and let the system be initialized so that, at time $t = 0$,

$$I = X = 20$$
$$Y = 12$$
$$S = 4 \text{ items/day}$$

```
TITLE   SIMPLE INVENTORY CONTROL SYSTEM
*
PARAM   K = (1.0, 0.5, 0.25, 0.125, 0.0625)
*
        S = 4.0 + 2.0*STEP(4.0)
        V = Y/T1
        U = S + K*(I - X)
        YDOT = U - V
        Y = INTGRL(Y0,YDOT)
        XDOT = V - S
        X = INTGRL(X0,XDOT)
*
CONST   I = 20.0, Y0 = 12.0, X0 = 20.0, T1 = 3.0
*
TIMER   DELT = 0.1, FINTIM = 50.0, PRDEL = 1.0
PRINT   U,V,X,Y
END
STOP
```

Figure 5-13. CSMP III program for an inventory control system.

Let the average delivery delay T_1 be 3 days. Suppose the demand suddenly increases at time $t = 4$ to 6 items/day. The changes in inventory level for various values of K are shown in Fig. 5-14. It can be seen that oscillations can occur. The ordering policy that has been adopted does, however, achieve its objective since the inventory settles to its planned level.

The oscillations are more pronounced when $K = 1$. They diminish as K decreases, and it can be shown that they disappear when $K = \frac{1}{12}$. It should be remembered that the term $K(I - X)$ has been added to the ordering *rate*. It is not

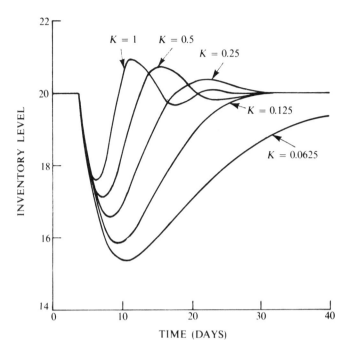

Figure 5-14. Inventory control system response.

a one-time order intended to replace a deficit detected in the inventory level. Some insight can then be gained as to why large values of K cause oscillations. With $K = 1$, for example, enough goods are shipped in a day to fill the deficit but, since there is a three-day delay in delivery, there is a long delay before the feedback begins to reduce the signal ordering supplies. The term added to S will then overfill the original deficit.

It would, of course, be possible to have an inventory control system based on the policy that, periodically, one-time orders are placed to wipe out accumulated inventory deficits. However, it is important to have some methodical way of correcting deviations automatically, particularly where demand fluctuates a great deal. The rule that has just been tested is a reasonably good rule provided that K is chosen to be a value for which there is little or no oscillation. A suitable value in this simple system is about $\frac{1}{2}$. The value will, of course, change with the delay time T_1. Many other inventory control rules could be used, and it requires a considerable amount of study to select a rule best suited for any particular system and type of demand, (6).

We have discussed this system in terms of a retailer selling goods. The basic problem of adjusting rates so that a certain level is maintained appears repeatedly in socio-economic systems, with the level representing less tangible items. The level might, for example, be some measure of the health of a population, with the coefficient K representing investment in medical services; or the level might be money supply, with K representing the influence of interest rates.

5-12

A Biological Example

As an example of how System Dynamics is applied in the biological field, consider the following example, (22). There are many examples in nature of parasites that must reproduce by infesting some host animal and, in so doing, kill the host. As a result, the population sizes of both the host and parasite fluctuate. As the parasite population grows, the host population declines. Ultimately, the decline in the number of hosts causes a decline in the birth of parasites and, consequently, the population of the hosts begins to climb. The process can continue to cause oscillations indefinitely.

To construct a model of the system, let X be the number of hosts and Y be the number of parasites. Suppose the excess of the birth rate of the hosts over the death rate from natural causes is A (assumed positive). In the absence of parasites, the population of the hosts should grow according to the equation

$$\dot{X} = AX$$

The death rate from infection by the parasites depends upon the number of encounters between the parasites and hosts, which is assumed to be proportional to the product of the numbers of parasites and hosts. The equation controlling the growth of the hosts is, therefore,

$$\dot{X} = AX - KXY$$

We will make the simplifying assumption that each death of a host due to a parasite results in the birth of one parasite. This is the only means by which the parasite population can grow, but the parasites are subject to a natural death rate of D. Hence the equation controlling the parasite population is

$$\dot{Y} = KXY - DY$$

Figure 5-15 shows a program for solving the two simultaneous equations for the particular values

$$A = 0.005$$
$$K = 6 \times 10^{-6}$$
$$D = 0.05$$

```
TITLE  PARASITE-HOST MODEL
*
       XDOT = A*X - K*X*Y
       YDOT = K*X*Y - D*Y
       X = INTGRL(X0,XDOT)
       Y = INTGRL(Y0,YDOT)
*
CONST  A = 0.005, D = 0.05, X0 = 10000.0, Y0 = 1000.0, K = 0.000006
*
TIMER  DELT = 0.1, FINTIM = 500.0, PRDEL = 10.0
PRINT  X,Y
END
STOP
```

Figure 5-15. Program for the host-parasite problem.

Figure 5-16 shows the output when the initial values of X and Y were 10,000 and 1,000, respectively. In this case, there are stable oscillations in both populations. It is possible for the system to be unstable, in which case either the parasite population dies out or both populations die. System Dynamic models of this type are used to find the range of conditions under which the populations remain in a stable

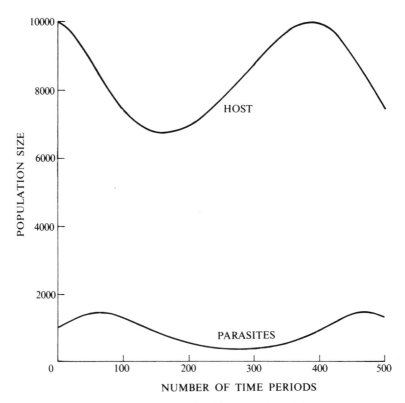

Figure 5-16. Results of host-parasite model.

oscillation, and determine what changes can throw the system into an unstable state with the parasite population dying out, (2) and (3). A study of that nature could establish what increase in the parasite death rate is needed to reach such a state, and so judge the scale of the necessary extermination campaign, (11).

5-13
World Models

The most widely known example of a System Dynamics model is the world model studied under the auspices of The Club of Rome, (18). There have, in fact, been several similar models produced by other investigators, as a result of the Club of Rome work, (1), (4), and (14). The Club of Rome is an informal, international group of individuals concerned with the future of mankind. They sponsored a study based on a model of the world, a model originally proposed by Professor Jay W. Forrester, (10). The results of the study have been reported, (16), and their generally gloomy tone has created a great deal of discussion.

The Club of Rome model investigates five aspects of global development: population growth, the spread of industrialization, pollution, the depletion of natural resources, and the incidence of malnutrition. The main conclusions drawn are that present trends place a limit on growth, the limit is likely to be reached in the next hundred years, and reaching that limit will result in a disastrous decline in both population and living conditions. It was also concluded that it is feasible to establish conditions that can lead to a state of equilibrium with satisfactory living conditions for everybody. (However, establishing these conditions requires an urgent cooperation among the nations of the world to implement some major changes in present practices.)

A typical section of the model uses a population growth model similar to that of Fig. 5-8 (b), but with both the birth and death rates affected by many factors. For example, the natural death rate is modified by the availability of food, the level of pollution, and the degree to which people are crowded as a result of industrialization. Each of these factors is influenced, directly or indirectly, by the existing population level, leading to a set of feedback loops, mostly of a positive type. Conflicts between industrialization and food production cause competition for capital and land, resulting in other forms of interaction between the variables of the model.

Inevitably, attempts to reflect the trends and their interactions in a mathematical model require some speculation and judgement, particularly in the matter of providing specific numbers for the model. Critics of the model have questioned the assumptions and data of the model; while advocates have pointed out the persistence of the main conclusions, with relatively small changes in the time scale, for a wide range of input values. Among the major topics of controversy has been the role of future technological changes, with the critics being unable to say what these will be, and the advocates finding it difficult to see feasible changes that would produce sufficient corrective action without changes in existing practices.

Some other work, based on input-output models rather than System Dynamics, however, has suggested ways in which adequate resources can be developed for future growth, (15).

5-14

The DYNAMO Language

Since a System Dynamics model consists principally of first order linear differential equations, any CSSL can be used to program a simulation run, (19). An example using CSMP III was given in Fig. 5-13. The results shown in Fig. 5-16 were also derived with a CSMP III program. A language that was developed especially for System Dynamics models (in their original implementation as Industrial Dynamics models) is called DYNAMO: an acronym for Dynamic Models. It is a particularly simple language designed for users who have little or no training in programming, (20).

Variables in DYNAMO are represented by symbols of from one to five characters, with some reserved names. The name TIME is reserved for making reference to the time in the system model. The way the passage of time is described is illustrated in Fig. 5-17. The instant at which calculations are being made is referred to as TIME.K. The previous instant at which calculations were made is TIME.J and the next instant at which calculations will be made is TIME.L. The interval just passed is called the JK interval, and the interval coming up is the KL interval. Calculations are made for uniform intervals of time, so that the JK and KL intervals

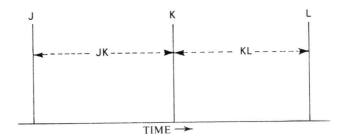

Figure 5-17. Representation of time in DYNAMO.

are always the same size. The length of the constant interval is designated by the symbol **DT**. The size of **DT** is chosen by the user.

The convention adopted is to attach one of the symbols J, K, L, JK, or KL as a suffix to a variable. The name of the variable is followed by the suffix separated by a period. Thus, a level called **INV** at time **J** will be indicated by **INV.J**, and the value of a rate **INP** during the interval **JK** will be indicated by **INP.JK**. Where a symbol is to represent a constant, it is not necessary to attach any suffix.

To simplify programming, DYNAMO defines a number of equation forms, each of which is a prototype. All equations must comply with these prototypes. The user selects the form of equation he desires to use, and completes the equation in accordance with the prototype structure, using the symbols of the particular variables to which the equation is applied. Each equation type defines a single variable on the left-hand side of the equation in terms of some combination of variables on the right-hand side. The equation form is numbered, and, if the variable it represents is a level or a rate, the number is followed by an **L** or **R**, respectively. Auxiliary variables can be defined by using the same equation forms and following the form numbers with the letter **A**. Initial values may also be entered with the same equation forms by using the letter **N**. A constant is entered by putting the letter **C** in column 1 and equating the symbol to its value in the statement field.

Some of the more important equation forms are listed in Table 5-2. The symbol **V** represents the variable being defined and the symbols **P**, **Q**, **R**, etc. represent other variables. Note that the first two equation forms are marked with an **L**, indicating that these may only be used to define levels. The others may be used for

TABLE 5-2 DYNAMO Equation Forms

Level Equations	Form Number	Exact Punching Format
$V = V + (DT)(P + Q)$	1L	V.K = V.J + (DT)(±P±Q)
$V = V + (DT)\dfrac{(P + Q)}{Y}$	3L	V.K = V.J + (DT)(1/±Y)(±P±Q)
$V = P$	6	V = ±P
$V = P + Q$	7	V = ±P±Q
$V = (P)(Q)$	12	V = (±P)(±Q)
$V = (P)(Q + R)$	18	V = (±P)(±Q±R)
$V = \dfrac{P}{Q}$	20	V = ±P/±Q
$V = \dfrac{P + Q}{R}$	21	V = (1/±R)(±P±Q)
Step Function P at time Q	45	V = STEP(±P, Q)
$V = \mathrm{MAX}(P, Q)$	56	V = MAX(±P, ±Q)
$V = \mathrm{MIN}(P, Q)$	54	V = MIN(±P, ±Q)

levels, rates, auxiliary variables, or initial conditions by following the number with L, R, A, or N, respectively.

The first level equation of Table 5-2 is the most commonly used method of integrating a differential equation. It is essentially carrying out the numerical computation method described in Sec. 3-6, and illustrated in Fig. 3-3. It projects the change in value of a variable over a small interval, by multiplying the rate of change of the variable by the interval size, and adding the result to the previous value of the variable in order to get the new value.

As an example of how coding is carried out, consider the flow chart of Fig. 5-18, in which a level INV depends upon a rate INP that increases the level and a rate OUT that decreases the level. Suppose the rates are proportional to two other levels LEV1 and LEV2 with the coefficients of proportionality being 1/T1 and 1/T2, respectively.

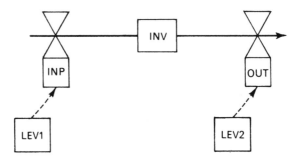

Figure 5-18. Coding of DYNAMO equations.

Algebraically, the equations can be written

$$INV = INV' + DT(INP - OUT)$$

$$INP = LEV1/T1$$

$$OUT = LEV2/T2$$

The first equation expresses the fact that the present level of INV is the level at the previous interval INV', modified by the changes brought about by the rates during the interval *DT*. It is a level equation of form **1L**. The other two equations are rate equations of form **20**. The DYNAMO coding is

```
1L      INV.K = INV.J + (DT)(INP.JK − OUT.JK)
20R     INP.KL = LEV1.K/T1
20R     OUT.KL = LEV2.K/T2
```

Exercises

5-1 Program the air-conditioner sales model of Sec. 5-9, allowing for break-downs. Assume the average time to sell a house is 5 months, the average time to install an air conditioner is 10 months, and breakdowns occur, on the average, after 25 months. Take the initial housing market to be 1,000 houses. Assume broken air conditioners are scrapped.

5-2 In the model of air conditioner sales, without breakdowns, that is given in Sec. 5-9, prove that the sales response cannot be oscillatory if K_1 and K_2 are positive. (Use the relation $(K_1 - K_2)^2 \geq 0$.)

5-3 Babies are born at the rate of 1 baby per annum for every 20 adults. After a delay of 6 years they reach school age. Their education takes 10 years, after which they are adults. The adults die after an average adult life of 50 years. Draw a System Dynamics diagram of the population and program the model, assuming the initial numbers of babies, school-children, and adults are, respectively, 300, 3,000, and 100,000.

5-4 A company establishes a pension fund for its employees by setting aside $\$P$ a month for each employee. The accumulated fund is invested and earns 5% per annum. The work force is expected to grow at 3% per annum. The company wants to study the soundness of its plan and proposes to simulate the effects of different assumptions about average length of service and average length of retirement. Draw a System Dynamics flow chart for the simulation.

5-5 The birth rate of a country is adding 100,000 people a year to an initial population of 5,000,000. The average life expectancy is 65 years. It is estimated that one ton of coal is consumed per annum for each individual. Draw a diagram showing how the country's resources of 500,000,000 tons are being depleted.

5-6 Two competing companies invest funds in capital equipment to improve their positions. The rate at which each invests funds decreases linearly as their own investment increases but increases linearly as their competitor's investment increases. Draw a diagram from which to simulate the competition and determine under what conditions the investments will stabilize.

5-7 The following model has been proposed to describe the population growth of a species of animal living in a confined space. The excess of births over natural deaths causes a growth rate of a times the current number of individuals, N. Competition for food causes death from starvation at a rate bN^2. Simulate the population growth assuming $a = 0.05$, $b = 1 \times 10^{-5}$, and $N = 500$ at zero time.

5-8 Assume in the previous problem that a second species exists in the same
space and is subject to the values $a_1 = 0.04$, $b_1 = 1.5 \times 10^{-5}$, and $N_1 =$
1,000 at time zero. However inter-species fighting causes deaths to both
species at a rate that is 1×10^{-6} times the product of their numbers. Simulate their population growths.

Bibliography

1 ATTINGER, E. O., ed., *Global System Dynamics*, New York: Halsted Press,
1970.

2 BARGER, I. A., P. R. BENYON, AND W. H. SOUTHCOTT, "Simulation of Larval
Population of a Parasitic Nematode of Sheep," *Simulation*, XXII, no. 1 (1974),
81–84.

3 BRAUER, F., A. C. SOUDACK, and H. S. JAROSCH, "Stabilization and De-
Stabilization of Predator-Prey Systems Under Harvesting and Nutriment
Enrichment," *Int, J. Control*, XXIII, no. 4 (1976), 553–573.

4 BURNETT, ROBERT A., AND PAUL J. DIONNE, "Globe 6: A Multiregion Interactive World Simulation," *Simulation*, XX, no. 6 (1973), 192–197 and 216ff.

5 CHADDA, ROSHAN L., AND SHARAD S. CHITGOPEKAR, "A 'Generalization' of
the Logistic Curves and Long-Range Forecasts of (1966–1991) Residence
Telephones," *Bell J. Econ. Manage. Sci.*, II, no. 2 (1971), 542–560.

6 CONWAY, RICHARD W., W. L. MAXWELL, AND L. W. MILLER, *Theory of
Scheduling*, Reading, Mass.: Addison-Wesley Publishing Company, Inc., 1967.

7 CROXTON, FREDERICK E., DUDLEY J. COWDEN, AND SIDNEY KLEIN, *Applied
General Statistics* (3rd ed.), Englewood Cliffs, N. J.: Prentice-Hall, Inc., 1967.

8 FORRESTER, JAY W., *Industrial Dynamics*, Cambridge, Mass.: The MIT Press,
1961.

9 ———, *Urban Dynamics*, Cambridge, Mass.: The MIT Press, 1969.

10 ———, *World Dynamics* (2nd ed.), Cambridge, Mass.: Wright-Allen Press,
Inc., 1973.

11 FREICHS, RALPH R., AND JUAN PRAWDA, "Computer Simulation Model for
the Control of Rabies in an Urban Area of Colombia," *Manage. Sci.*, XXII,
no. 4 (1975), 411–421.

12 HAMILTON, HENRY R., ed., *System Simulation for Regional Analysis*, Cambridge, Mass.: The MIT Press, 1969.

13 JAQUETTE, DAVID L., "Mathematical Models for Controlling Growing
Populations: A Survey," *Oper. Res.*, XX, no. 6 (1972), 1142–1151.

14 KRENZ, JERROLD H., "World Dynamics — An Alternative Model," *IEEE Trans. Syst. Cybern.*, vol. SM-3, no. 3 (1973), 272–275.

15 LEONTIEF, WASSILY W., et al., *The Future of the World Economy*, New York: UN Dept. of Economic and Social Affairs, 1976.

16 MEADOWS, DENNIS L., et al., *Dynamics of Growth in a Finite World*, Cambridge, Mass.: Wright-Allen Press, Inc., 1974.

17 MEIER, ROBERT C., WILLIAM T. NEWELL, AND HAROLD L. PAZER, *Simulation in Business and Economics*, pp. 86–92, Englewood Cliffs, N. J.: Prentice-Hall, Inc., 1969.

18 PECCEI, AURELIO, "The Club of Rome — The New Threshold," *Simulation*, XX, no. 6 (1973), 199–206.

19 PETERSON, NORMAN D., "MIMIC, An Alternative Programming Language for Industrial Dynamics, *Socio-Economic Planning Sciences*, VI, no. 3 (1972), 319–327.

20 PUGH, ALEXANDER, III, *DYNAMO II User's Manual*, Cambridge, Mass.: The MIT Press, 1973.

21 *United Nations Statistical Year Book, 1973*, Table 175, New York: The United Nations, 1974.

22 VOLTERRA, VITO, *Leçons sur la Théorie Mathematique de la Lutte pour la Vie*, Paris: Gauthier-Villars, et Cie., 1931.

6

PROBABILITY CONCEPTS
IN SIMULATION

6-1

Stochastic Variables

As was mentioned in Sec. 1-3, when the outcome of an activity can be described completely in terms of its input, the activity is said to be deterministic. However, there are many activities where the output is random. Activities, or processes, of this nature are said to be stochastic. The terms *random* and *stochastic* are used interchangeably. More formally, a *stochastic process* is defined as being an ordered set of random variables: the ordering of the set usually being with respect to time. In particular, stochastic activities used in system simulation give rise to a stochastic variable, represented by a sequence of random numbers occurring over time.

The activity can be discrete or continuous. The stochastic variable might, for example, represent the number of days in the month (or a sequence of dates), that take only the values 28, 29, 30, or 31; or it might be a variable representing wind velocity, which can vary over a continuous range.

In order to conduct simulation studies, it is necessary to be familiar with the ways of describing stochastic variables, and of generating random numbers to represent them. We collect together in this chapter most of the definitions and methods needed for these purposes. The next chapter will also discuss some special probability functions, used for specific purposes.

6-2

Discrete Probability Functions

If a stochastic variable can take I different values, x_i ($i = 1, 2, \ldots, I$), and the probability of the value x_i being taken is $p(x_i)$, the set of numbers $p(x_i)$ is said to be a *probability mass function*. Since the variable must take one of the values, it follows that $\sum_{i=1}^{I} p(x_i) = 1$. Examples of discrete variables that can occur in a simulation study are the number of items a customer buys in a store, or the origin of a message in a communication system where the terminals creating messages have been numbered from 1 to n.

The probability mass function may be a known set of numbers. For example, with a die, the probability of each of the six faces is $1/6$, but, frequently, a probability mass function must be estimated by counting how many times each possible value occurs in a sample. The probability is then estimated by computing the fraction of all observations that took each value.

Table 6-1, for example, gives data gathered on the number of items bought in a store for a sample of 250 customers. The number of items, x_i, is shown in the

TABLE 6-1 Number of Items Bought by Customers

Number of Items x_i	Number of Customers n_i	Probability Distribution $p(x_i)$	Cumulative Distribution $P(x_i)$
1	25	0.10	0.10
2	128	0.51	0.61
3	47	0.19	0.80
4	38	0.15	0.95
5	12	0.05	1.00
	250		

first column. The second column is the number of customers who bought that number of items, n_i. The third column is the estimate of the probability of buying x_i items, derived by dividing the number of customers that bought that many items by the total number of observations, N. That is,

$$p(x_i) = \frac{n_i}{N} \tag{6-1}$$

A *cumulative distribution function*[1] is defined as a function that gives the probability of a random variable's being less than or equal to a given value. In the case of discrete data, the cumulative distribution function is denoted by $P(x_i)$. The values in column four of Table 6-1, which accumulate the values in column three, are estimates of the cumulative distribution function. From its definition, a cumulative distribution function cannot decrease with increasing values of the random variable, and it must increase to a maximum value of 1.

Probability functions are often displayed graphically. Figure 6-1 shows the data of Table 6-1 displayed in graphic form. On the left, Fig. 6-1(a), is the probability mass function, represented by the figures in column three of Table 6-1. On the right, Fig. 6-1(b), is the cumulative distribution function, given by the figures in column four of Table 6-1.

The probability mass function appears as a bar histogram. The cumulative distribution function appears as a step function type of graph. It is necessary to decide how the discontinuities of the graph are to be interpreted. In Fig. 6-1(b), the blobs and dashed lines are meant to indicate that the value of the cumulative dis-

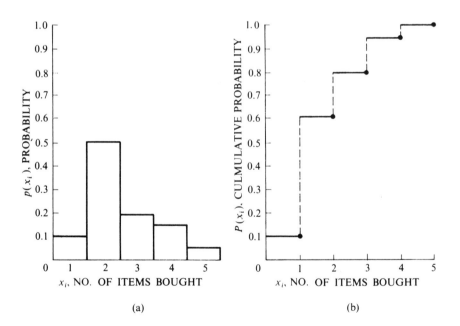

Figure 6-1.   Discrete probability functions.

[1]In the mathematical literature, this function is simply called the distribution function. It is regarded as the fundamental description of a random process, from which the probability density function is derived—rather than the other way around, as has been implied in the development of this text. The approach and terminology used here is adopted as being more descriptive of the way data are derived in practice.

tribution at the point of discontinuity is to be associated with the lower value of x_i. Thus, the figure of 0.61 is associated with two items being bought, and not three. In an actual system, the question of where to place the precise value of 0.61 is not likely to be important, but, in the numerical computations of a system simulation, it is important to be unambiguous and consistent.

6-3
Continuous Probability Functions

When the variable being observed is continuous and not limited to discrete values, an infinite number of possible values can be assumed by the variable. The probability of any one specific value occurring must logically be considered to be zero. To describe the variable, a *probability density function*, $f(x)$, is defined. The probability that x falls in the range x_1 to x_2 is given by

$$\int_{x_1}^{x_2} f(x)\, dx \qquad f(x) \geq 0$$

Intrinsic in the definition of a probability density function is the property that the integral of the probability density function taken over all possible values is 1. That is,

$$\int_{-\infty}^{\infty} f(x)\, dx = 1$$

The lower limit of the integral is shown as being $-\infty$. In practice, most variables in a simulation study have a finite lower limit, generally zero. The probability density function at and below this limit is then identically zero, and the lower limit of the integral can be replaced by the finite value. The same effect may occur at the upper limit.

The cumulative distribution function, which defines the probability that the random variable is less than or equal to x, is denoted by $F(x)$, in the case of a continuous variable. It is related to the probability density function, as follows:

$$F(x) = \int_{-\infty}^{x} f(x)\, dx$$

From its definition, $F(x)$ is a positive number ranging from 0 to 1, and the probability of x falling in the range x_1 to x_2 is $F(x_2) - F(x_1)$. Figure 6-2 illustrates a probability density function. The equation for the curve is

$$f(x) = xe^{-x} \qquad x \geq 0 \qquad (6-2)$$

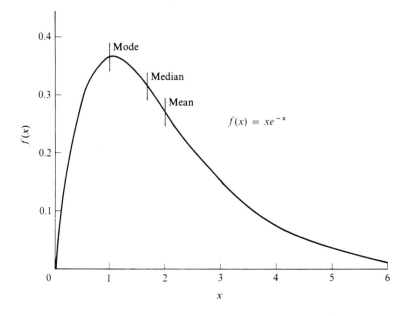

Figure 6-2. A probability density function.

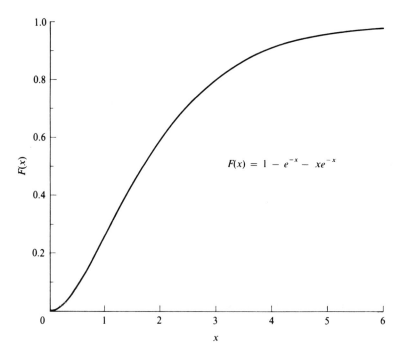

Figure 6-3. A cumulative distribution function.

(The points of the curve that have been marked will be described shortly.) In this case, the density function is defined only for non-negative values of x, but it extends for all positive values of x.

The function can be integrated to give the following equation for the cumulative distribution:

$$F(x) = 1 - e^{-x} - xe^{-x}$$

The cumulative distribution is plotted in Fig. 6-3. This also is defined for all positive values of x, and, as is required for a cumulative distribution function, it never decreases with increasing x, and it tends to the limiting value of 1.

6-4

Measures of Probability Functions

Probability functions are summarized by certain key parameters. The most important of these is the *mean value*, denoted by m.[2] In the case of a discrete distribution, if there are N observations, taking the individual values $x_r (r = 1, 2, \ldots, N)$, the mean is estimated from the formula

$$m = \frac{1}{N} \sum_{r=1}^{N} x_r \qquad (6\text{-}3)$$

If the observations fall into I groups, where the ith group takes the value x_i and has n_i members, the formula can be written as

$$m = \frac{1}{N} \sum_{i=1}^{I} n_i x_i$$

For the data of Table 6-1, for example, there are five groups ($I = 5$); the values of x_i are 1, 2, $\ldots$, 5; and the numbers n_i are 25, 128, 47, 38, and 12.

Taking the term N inside the summation sign, and recalling Eq. (6-1), this may be written as

$$m = \sum_{i=1}^{I} p(x_i) x_i$$

For the data of Table 6-1, it will be found that the formula leads to the mean value of 2.54.

[2]The true values of probability functions measures are usually denoted by Greek letters. Roman letters are used to indicate estimates of the true values. The distinction is not important here. Roman notation is therefore used, since the prime interest is in estimation.

For a continuous variable, the mean value is defined by the following integral:

$$m = \int_{-\infty}^{\infty} xf(x)dx \qquad (6\text{-}4)$$

When $f(x)$ is specified by a mathematical function, it may be possible to carry out the evaluation of the mean analytically. Otherwise, the integration must be carried out numerically, as described in the next section. In the case of the probability density function plotted in Fig. 6-2, it is possible to carry out the integration analytically, leading to the value of 2, which is marked as being the mean in Fig. 6-2.

The mean value is also called the *expected value*, a term that describes more fully the nature of a mean value. If a single number has to be picked to represent the variable without any randomness, the mean, or expected, value is the best one to use—even though, in the case of a discrete variable, the mean value can be a number that cannot actually occur. This happened in the calculation just made for the mean number of items bought in a store, where the value was 2.54. In fact, it is unlikely that the mean value of a set of discrete data will coincide with one of the feasible values.

If the random value is being replaced by a single number, and it is important that a feasible number be used, the best choice is the nearest feasible number to the mean value, in the present case, 3. Such rationalizing, however, should not be made until it is absolutely necessary. If, for example, an estimate has to be made of how many times an item will have to be rung on the cash register of the store, it is far more accurate to multiple the mean number of items per customer, 2.54, by the expected number of customers, which, as the result of a similar calculation, is probably another non-integral number, and then take the nearest integer. To rationalize the individual mean values to integers before multiplying could produce a substantial error.

There are two other measures of a distribution that must not be confused with the mean. These are the mode and the median. The mean is not necessarily the most frequently occurring value of a random variable. When the probability density function has a peak, the value of x at which the peak occurs is called a *mode*. If there is only one peak, as occurs in the function of Fig. 6-2, the probability density function is said to be unimodal. The mode is then the most frequently occurring value. By differentiating Eq. (6-2) for $f(x)$ and equating to zero, it will be found that the maximum occurs at $x = 1$, and that value is marked in Fig. 6-2 as being the mode. The probability density function can have more than one mode, in which case the most frequently occurring value is the mode that has the highest peak.

It is not even necessarily true that half the values of a random variable will fall below the mean value. The value of x which defines this point is called the *median*. It is easily found from the cumulative distribution, since it is the point at which $F(x) = 0.5$. In the case of the function of Fig. 6-2, the median falls at $x = 1.68$, which is marked as such.

The probability density function is sometimes described in terms of *fractiles*, which are a generalization of the median. Ten fractiles, for example, may be chosen to describe the data of the first tenth, the second tenth, and so on. When considering percentages, the fractiles are called *percentiles*. They can easily be found from the cumulative distribution function.

In Fig. 6-2, the mean happens to fall at a point greater than the mode, and the median falls between the two. However, there is no general relation between the order of the three measures. If the probability density function is symmetric, so that an axis can be found to divide the curve into two mirror images, then the mean and median coincide. If the function is unimodal, the mode will also coincide with the same value. The mean value involves weighting the data according to their positions when the data are ordered by magnitude; the median depends only on the order; and a mode depends only on magnitude. The effect of one more observation, therefore, affects the three measures differently. A relatively infrequently occurring value may carry a large weight when calculating the mean if it occurs at a position far removed from the other observations. The same point will have a relatively small effect on the median, and will most likely have no effect on the mode. When the mean is seen to be far removed from the median or mode, care should be taken to check that erroneously recorded values are not distorting the results.

In general, however, the mode and the median are not particularly significant when representing probability functions in a system simulation, but it is important not to confuse them with the mean when interpreting input data. In addition, when certain well-known functions are to be fitted to experimental data, the calculations determining the best fit sometimes involve the use of modes, medians, or fractiles. See, for example, the method of fitting the lognormal function, given in (3).

Another measure of a stochastic variable that *is* important is the variance of the variable, denoted by s^2. The practical form in which the variance is used is as a *standard deviation*, which is the (positive) square root of the variance, denoted by s. If there are N observations taking the values x_r ($r = 1, 2, \ldots, N$), and having the mean value m, the standard deviation is estimated with the formula

$$s = \left\{ \frac{1}{(N-1)} \sum_{r=1}^{N} (m - x_r)^2 \right\}^{1/2} \tag{6-5}$$

Expanding the terms in parentheses under the summation sign, and recalling the definition of the mean, given in Eq. (6-3), this may be written as

$$s = \left\{ \frac{1}{(N-1)} \left(\sum_{r=1}^{N} x_r^2 - Nm^2 \right) \right\}^{1/2} \tag{6-6}$$

The summations of Eqs. (6-5) and (6-6) are carried over the individual observations. The calculations can be simplified when the observations fall into groups of the same value. If there are I groups, where the ith group takes the value x_i and has n_i members, the formula of Eq. (6-5) can be written

$$s = \left\{ \frac{1}{(N-1)} \sum_{i=1}^{I} n_i(m - x_i)^2 \right\}^{1/2}$$

If the difference between N and $N - 1$[3] is ignored, the ratios $n_i/(N - 1)$ that then occur can be replaced by $p(x_i)$, in accordance with Eq. (6-1), and the formula becomes

$$s = \left\{ \sum_{i=1}^{I} p(x_i)(m - x_i)^2 \right\}^{1/2}$$

Since $\sum_{i=1}^{I} p(x_i) = 1$ and $\sum_{i=1}^{I} p(x_i)x_i = m$, the formula becomes

$$s = \sum_{i=1}^{I} p(x_i)x_i^2 - m^2 \tag{6-7}$$

The definition of the standard deviation for a continuous variable is closely related to the formula of Eq. (6-7). If the number of groups is increased indefinitely, the probability mass function, $p(x_i)$ tends to the probability density function, $f(x)$, so that, in the limit,

$$s = \int_{-\infty}^{\infty} f(x)x^2 \, dx - m^2 \tag{6-8}$$

where m is defined by Eq. (6-4). In the case of the function of Fig. 6-2, for example, it will be found that the standard deviation is 1.414.

The significance of the standard deviation is that it is a measure of the degree to which the data are dispersed. This is best seen from Eq. (6-5) which defines the standard deviation for discrete data in terms of the individual observations. Because of the squared terms occurring under the summation sign, every observation that does not exactly coincide with the mean value contributes a positive amount to the sum defining the standard deviation. The farther the observation deviates from the mean, irrespective of whether the deviation is positive or negative, the bigger is the contribution. Points that coincide with the mean value make zero contribution. The size of the standard deviation is, therefore, a measure of the degree to which data are dispersed from the mean value.

[3]The slight bias introduced by doing so can be removed by multiplying the result by $N/(N-1)$.

Because of the square root occurring in the definition, the standard deviation has the same dimensionality as the observations. It can therefore be compared directly with the mean value. The ratio of the standard deviation to the mean value, which is called the *coefficient of variation*, is a more useful measure of dispersion, since it is expressed in relative terms, and does not depend upon the scale in which the data are described.

The formulae on which the estimation of a standard deviation is based assume that the observations are mutually independent; that is to say, the value observed at any time is not related in any statistically significant way with any previously observed value, or values, other than the fact that they are randomly drawn from a population with the same probability density function. Data that do not meet this criterion are said to be autocorrelated. An example may help to clarify this concept.

Suppose the times people spend on a waiting line are being measured. Assuming a simple first-come, first-served discipline, any time there are two or more people waiting, the time the second person waits cannot be less then the time the first person waits, and so on down the line. Each of these times is not independent of the previous time. The data are autocorrelated and the formulae given above cannot be applied.[4]

By considering the process that is generating the data, (in particular, looking for cyclic effects, which always imply autocorrelation) it may be apparent that autocorrelation is present. Since measuring and correcting autocorrelation effects arise in connection with interpreting simulation results, these matters will be discussed later, in Chap. 14, which is concerned with the interpretation of simulation results.

In addition, the formulae given for estimating both the mean and standard deviation assume that the distribution from which observations are drawn is stationary, that is, is independent of the time at which observations are made. Many systems have obvious violations of this condition, as when the system load depends upon the time of day. Care must be taken to see that measurements are made over consistent periods of time in which it is reasonable to suppose the distribution is stationary. Where variations do occur with time of day, a first, usually reasonable, assumption is that the output of a random-number generator can be scaled to reflect the change in mean value. However, the effect this has on the standard deviation cannot be controlled. The method should not be used when the variations with time are too large. Instead, a series of distributions will need to be used, as time increases.

---

[4]The formula for estimating the mean, however, remains valid, (4).

6-5
Numerical Evaluation of
Continuous Probability Functions

In practice, most probability functions used in system simulation are derived from experimental data. Certain processes may be represented by mathematical functions derived on theoretical grounds. The next chapter, describing arrival patterns and service times, will give some examples. It is also possible that the experimental data can be fitted by a mathematical function. Unless it appears that the fitted function is of a particularly simple form, or can be taken as one of the theoretical functions mentioned in the next chapter, there is no particular merit in fitting a function to the data. The purpose of introducing a probability function into a simulation is to generate random numbers with that particular distribution. As will be shown shortly, the principal way of generating such numbers involves mathematical operations that usually cannot be carried out analytically. Numerical techniques will, therefore, have to be used, so the data might as well be left in numerical form from the beginning.

The customary way of organizing data derived from observations is to display them as a *frequency distribution*, which shows the number of times the variable falls in different intervals. For example, suppose that 1,000 observations of telephone call lengths are made and tabulated in intervals of 10 seconds. The frequency distribution might then appear as shown in Table 6-2. The leftmost column defines a number of intervals, and the next column records the number of calls whose length fell within that interval. The distribution can also be displayed graphically in a bar-graph, as shown in Fig. 6-4. (Some of the values included in Table 6-2 are too small to show in the figure.)

A more useful way of describing the same information is as a *relative frequency distribution*, in which the number of observations for each interval is divided by the total number of observations. Table 6-2 shows the relative frequency in the third column and Fig. 6-4 displays the relative frequency on the axis to the right of the figure.

From the definition, it is apparent that the sum of all the values of the relative frequency distribution is 1. It is important to note, however, that the relative frequency distribution is *not* necessarily the probability density function for the variable. If the n intervals for which the frequency distribution has been generated are $x_i < x \leq x_{i+1}$ ($i = 0, 1, \ldots, n-1$), and p_i is the relative frequency count for the ith interval, then p_i must be interpreted as the integral of the probability density function over the interval; that is,

$$p_i = \int_{x_i}^{x_{i+1}} f(x)\, dx$$

TABLE 6-2 Distribution of Telephone Call Lengths

Call Length (sec)	Number of Calls	Relative Frequency	Probability Density	Cumulative Distribution
0	0	0.000	0.0000	0.000
10	1	0.001	0.0001	0.001
20	1	0.001	0.0001	0.002
30	1	0.001	0.0001	0.003
40	2	0.002	0.0002	0.005
50	8	0.008	0.0008	0.013
60	28	0.028	0.0028	0.041
70	65	0.065	0.0065	0.106
80	121	0.121	0.0121	0.227
90	175	0.175	0.0175	0.402
100	197	0.197	0.0197	0.599
110	175	0.175	0.0175	0.774
120	121	0.121	0.0121	0.895
130	65	0.065	0.0065	0.960
140	28	0.028	0.0028	0.988
150	9	0.009	0.0009	0.997
160	2	0.002	0.0002	0.999
170	1	0.001	0.0001	1.000

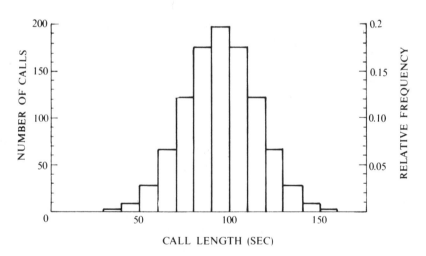

Figure 6-4. Frequency distribution of telephone call lengths.

The choice of the interval size will normally be made on the assumption that it is small enough to ignore variations of the probability density function over the interval. If so, the value of the density function in each interval is $p_i/(x_{i+1} - x_i)$. For the case of the data in Table 6-2, for example, the relative frequency distribu-

tion for each interval is divided by 10 to get the approximation to the probability
density function shown in the table as the fourth column. If the intervals for which
the observations were taken are not all the same size, appropriate adjustments must
be made.

In practice, there is more interest in deriving the cumulative distribution function. If the values of p_i are accumulated, the successive values, $F_r = \sum_{i=0}^{r} p_i$
$(r = 1, 2, \ldots, n)$, represent the value of the cumulative distribution function at
the points $x_r (r = 1, 2, \ldots, n)$. This is shown as the fifth column of Table 6-2.
A series of straight line segments drawn through these points, as illustrated in
Fig. 6-5, can be taken as an approximation to the cumulative distribution function.
The straight line approximation conforms with the previous assumption that the
probability density function is constant over each interval. A better approximation
can be attempted by fitting a smooth curve to the points; in which case, the probability density function can be derived by plotting the slope of the cumulative
distribution curve.

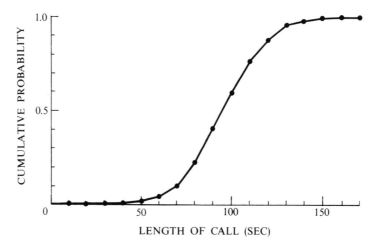

Figure 6-5. Cumulative distribution of telephone call lengths.

Probability functions have been discussed so far as a means of describing the
behavior of a stochastic variable. They record the results of a number of observations, considering only the values assumed by the variable and not the sequence
in which they occurred. In carrying out a simulation that involves stochastic variables, the reverse problem arises. It is necessary to generate a sequence of numbers
in which the successive values are random and have the distribution that describes
the stochastic variable. The continuous uniform distribution, discussed in the next
section, is an important example.

6-6

Continuous Uniformly Distributed
Random Numbers

By a continuous uniform distribution we mean that the probability of a variable, X, falling in any interval within a certain range of values is proportional to the ratio of the interval size to the range; that is, every point in the range is equally likely to be chosen. Suppose the possible range of values is from A to B ($B > A$), then the probability that X will fall in an interval Δx is $\Delta x/(B - A)$. Drawn as a graph, the probability density function is a straight line of height $1/(B - A)$ between the points A and B, as shown in Fig. 6-6 (a).

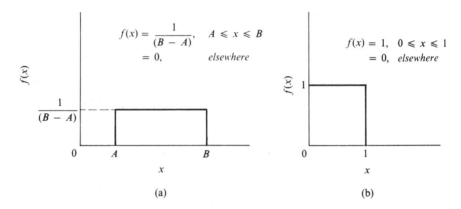

(a)   (b)

Figure 6-6. Continuous uniform distributions.

There is no loss in generality in assuming that the range is from 0 to 1, because, if X_i is a sequence of uniformly distributed numbers in the range 0 to 1, then $(B - A)X_i + A$ is a uniformly distributed sequence in the range A to B. The following discussion will therefore consider the case of the range being from 0 to 1, so that the probability density function is, as illustrated in Fig. 6-6 (b),

$$f(x) = 1 \qquad 0 \le x \le 1$$
$$= 0 \qquad \text{elsewhere}$$

In dealing with numerical calculations, a certain number of digits are chosen to represent a quantity, depending upon the desired accuracy. Strictly speaking, it is not then possible to represent a continuous variable, since the finite number of digits allows only a finite number of possible values. Given enough digits, the number of possible values that can be assumed is sufficiently large to treat the variable as being continuous. It will be assumed that the generated random numbers are represented by sufficient digits to make the "granularity" caused by the finite number of digits negligible. Caution must be used, however, if a random number is modified subsequently to its generation. If the number is multiplied by a factor that is greater than 1, there is an increase in "granularity" which can lead to a loss of accuracy. Most digital computer programs using a random number generator use a full computer word to represent the random number, so that the number has sufficient binary digits. However, for accurate calculations, it is not unusual to use double precision for the calculation of random numbers.

There are many physical processes that can be considered to give a sequence of uniformly distributed random numbers. For example, an electrical pulse generator can be made to drive a counter cycling from 0 through 9. Using an electronic noise generator or a radioactive source, the pulses can be generated at random intervals. By sampling the counter at regular intervals, the value can be taken as a digit of the desired random number. Repeating the process many times, or running several counters in parallel, a random number of any desired number of digits can be created, (1).

Tables of uniformly distributed random numbers produced by such physical processes have been published. Reference (12) is a particularly comprehensive set of numbers generated by a method based on the process just described. Table 6-3 is a set of random numbers reproduced from a single page of Ref. (12). The digits have been arranged in columns of five, and the columns and rows have been further divided into groups for ease of reading.

Suppose it is decided to generate a series of uniformly distributed random numbers between 0 and 1 to an accuracy of 1 part in 10,000; that is, the numbers are to have 4 decimal digits. The first column on the left can be read row-by-row discarding, say, the last digit, and when the column is finished the next column can be used. Using this method, we find the first five random numbers are

$$0.1009$$
$$0.3754$$
$$0.0842$$
$$0.9901$$
$$0.1280$$

If numbers with more than five digits are needed, columns are combined. It is not necessary to start with the first column. In fact, for proper use, particularly when a calculation is being repeated with different sets of random numbers, the starting points should be chosen at random. This could be done by using a random number to decide upon the page, column, and row at which to start.

TABLE 6-3 Table of Random Digits

10097	32533	76520	13586	34673	54876	80959	09117	39292	74945
37542	04805	64894	74296	24805	24037	20636	10402	00822	91655
08422	68953	19645	09303	23209	02560	15953	34764	35080	33606
99019	02529	09376	70715	38311	31165	88676	74397	04436	27659
12807	99970	80157	36147	64032	36653	98951	16877	12171	76833
66065	74717	34072	76850	36697	36170	65813	39885	11199	29170
31060	10805	45571	82406	35303	42614	86799	07439	23403	09732
85269	77602	02051	65692	68665	74818	73053	85247	18623	88579
63573	32135	05325	47048	90553	57548	28468	28709	83491	25624
73796	45753	03529	64778	35808	34282	60935	20344	35273	88435
98520	17767	14905	68607	22109	40558	60970	93433	50500	73998
11805	05431	39808	27732	50725	68248	29405	24201	52775	67851
83452	99634	06288	98083	13746	70078	18475	40610	68711	77817
88685	40200	86507	58401	36766	67951	90364	76493	29609	11062
99594	67348	87517	64969	91826	08928	93785	61368	23478	34113
65481	17674	17468	50950	58047	76974	73039	57186	40218	16544
80124	35635	17727	08015	45318	22374	21115	78253	14385	53763
74350	99817	77402	77214	43236	00210	45521	64237	96286	02655
69916	26803	66252	29148	36936	87203	76621	13990	94400	56418
09893	20505	14225	68514	46427	56788	96297	78822	54382	14598
91499	14523	68479	27686	46162	83554	94750	89923	37089	20048
80336	94598	26940	36858	70297	34135	53140	33340	42050	82341
44104	81949	85157	47954	32979	26575	57600	40881	22222	06413
12550	73742	11100	02040	12860	74697	96644	89439	28707	25815
63606	49329	16505	34484	20419	52563	43651	77082	07207	31790
61196	90446	26457	47774	51924	33729	65394	59593	42582	60527
15474	45266	95270	79953	59367	83848	82396	10118	33211	59466
94557	28573	67897	54387	54622	44431	91190	42592	92927	45973
42481	16213	97344	08721	16868	48767	03071	12059	25701	46670
23523	78317	73208	89837	68935	91416	26252	29663	05522	82562
04493	52494	75248	33824	45862	51025	61962	79335	65337	12472
00549	97654	64051	88159	96119	63896	54692	82391	23287	29529
35963	15307	26898	09354	33351	35462	77974	50024	90103	39333
59808	08391	45427	26842	83609	49700	13021	24892	78565	20106
46058	85236	01390	92286	77281	44077	93910	83647	70617	42941
32179	00597	87379	25241	05567	07007	86743	17157	85394	11838
69234	61406	20117	45204	15956	60000	18743	92423	97118	96338
19565	41430	01758	75379	40419	21585	66674	36806	84962	85207
45155	14938	19476	07246	43667	94543	59047	90033	20826	69541
94864	31994	36168	10851	34888	81553	01540	35456	05014	51176
98086	24826	45240	28404	44999	08896	39094	73407	35441	31880
33185	16232	41941	50949	89435	48481	88695	41994	37548	73043
80951	00406	96382	70774	20151	23387	25016	25298	94624	61171
79752	49140	71961	28296	69861	02591	74852	20539	00387	59579
18633	32537	98145	06571	31010	24674	05455	61427	77938	91936
74029	43902	77557	32270	97790	17119	52527	58021	80814	51748
54178	45611	80993	37143	05335	12969	56127	19255	26040	90324
11664	49883	52079	84827	59381	71539	09973	33440	88461	23356
48324	77928	31249	64710	12295	36870	32307	57546	15020	09994
69074	94138	87637	91976	35584	04401	10518	21615	01848	76938

6-7

Computer Generation of Random Numbers

It is possible to read a table of numbers, such as Table 6-3, into a computer, but the required quantity of random numbers can be very large and it is very inconvenient to prepare data in this form. As a result, computer programs have been developed for generating sequences of numbers uniformly distributed over a given range (usually 0 to 1). Starting with an initial number, the methods most commonly used specify a procedure for generating a second number. Using the second number as input, the procedure is repeated to produce a third number, and so on. In general, the $(i + 1)$th number of the sequence is generated from the ith number. An extensive literature exists on the generation of random numbers, (11).

The particular method that has come to be most widely used is called the congruence method or, sometimes, the residue method, (9). Given three constants λ, μ, and P, the procedure derives the $(i + 1)$th number from the ith number by multiplying by λ, adding μ, and then taking the remainder, or residue, upon dividing by P; this procedure is described mathematically by the expression

$$c_{i+1} = (\lambda c_i + \mu) \qquad \text{(modulo } P\text{)}$$

To begin the process, an initial number c_0 is needed, and this is called a seed. Complete specification of the process therefore needs four numbers, λ, μ, c_0, and P.

It is apparent that it is impossible to produce a nonrepeating sequence of numbers from the procedure. The set of numbers represented by n digits is finite, so ultimately some member of the sequence repeats an earlier value and the sequence is repeated thereafter. However, careful selection of the constants results in a very long sequence before such a repetition occurs, so that for practical purposes the sequence can be said to produce random numbers. Because of the ultimate repetition of the sequence, the term *pseudo-random number generator* is used to describe the procedure. Henceforth, the unqualified use of the term random number will mean a uniformly distributed random number drawn from a pseudo-random number generator. The symbol U will be used to denote such a random number.

Three types of congruence pseudo-random number generators have been used. They are additive, multiplicative, and mixed congruential generators. A mixed method is defined by the formula that was just given. If $\lambda = 1$, the method is said to be additive, and, if $\mu = 0$, the method is said to be multiplicative. Some studies (2) have indicated that the multiplicative method is superior to the additive method and that mixed methods are not noticeably better than simple multiplicative methods. Considering this form of the generation method, the formula defining the method is

$$c_{i+1} = \lambda c_i \quad \text{(modulo } P\text{)}$$

The problem of choosing good values for the constants of a pseudo-random number generator is a complex one. To be considered random, the sequence of numbers produced by the generator must meet various tests to ensure that they are uniformly distributed, and that there are no significant correlations between the digits of individual numbers or between sequential numbers. At the same time, it is important to ensure that the length of the cycle before the sequence of numbers begins to repeat is sufficiently long. A practical consideration is that the numbers can be computed efficiently on a digital computer.

It is not possible, in general, to predict how a generator based on a particular set of numbers will perform, other than in certain trivial cases. In the search for good generators various theoretical results have been produced to indicate what conditions are likely or unlikely to produce good generators. Using these results, the range of choice can be narrowed, but the final choice must be tested. In particular, the choice depends upon the word size of the computer to be used. Values selected as being good for one type of computer cannot automatically be adopted for use on another.

It is tempting to select the number P to be a power of the number base being used by a computer—that is, in the case of binary computers to make $P = 2^n$, where n is an integer. The reason is that the operation of dividing by P, needed to carry out the modulo operation, is simple, because it only involves moving the binary point by a shift operation. It has been shown (5), however, that this inevitably leaves certain gaps in the spectrum of possible numbers.

A set of guidelines that have been proposed (6) is that P be chosen as the largest prime that can be fitted to the computer word size, and that λ be a positive primitive root of P.[5] A generator based on these principles which is suitable for a 35-bit machine is discussed in (7). It uses $P = 2^{35} - 31$ and $\lambda = 5^5$. Another generator of this nature, based on a 31-bit word size,[6] is described in (10). That generator uses $P = 2^{31} - 1$, which happens to be a prime number and is simultaneously the largest number that can be represented in 31 bits. Tests for various values of λ were made. The one selected is $\lambda = 16807$, which is a positive primitive root of the form 7^5.

---

[5] A positive primitive root is defined as a prime factor or any positive integral power thereof.

[6] The counts apply to bits assigned as numerical digits. With the addition of a sign bit, the first case applies to a 36-bit word machine, and the second to a 32-bit word machine.

There is no general consensus on what constitutes a necessary and sufficient set of statistical tests for a random number generator. The reader is referred to (10), which describes the various tests and, in particular, discusses the results of the tests applied to the 31-bit word generator that has just been described.

6-8

A Uniform Random Number Generator

Although the number of computational steps involved in a congruence method of generating a random number is small, so that no great penalty is paid for repeating the coding wherever a random number is needed, it is customary to create a subroutine that can be used as a common source of all random numbers.

There is often a need for both floating-point and integer random numbers, and generators are sometimes written to produce both forms of number.

Table 6-4 shows a computer program that implements the multiplicative congruence generator, based on the 31-bit word size, that was described in the last section. It is written in System/360 Basic Assembler Language. It can be used by any program that conforms to the System/360 or 370 FORTRAN linkage conven-

TABLE 6-4   A Program for a Pseudo-Random Number Generator[7]

```
RANDOM    CSECT
          USING   *,15              INITIAL LINKAGE
          STM     2,5,28(13)
          LM      2,3,0(1)          LOAD ADDRESSES OF
                                    VARIABLES PASSED
          L       5,A               COMPUTE NEXT INTEGER
          M       4,0(2)            RANDOM NUMBER WITH
                                    X(I + 1) = AX(I) (MOD P)
          D       4,P
          ST      4,0(2)
          SRL     4,7               COMPUTE NEXT REAL
                                    RANDOM NUMBER
          A       4,CHAR
          ST      4,0(3)
          LM      2,5,28(13)        TERMINAL LINKAGE
          BR      14
CHAR      DC      F'1073741824'     CONSTANTS. CHAR FIRST
A         DC      F'16807'          SO A IS ON DOUBLE WORD
P         DC      F'2147483647'     BOUNDARY. MAKES LM
          END                       INSTRUCTION FASTER.
```

[7]Reprinted by permission from the IBM Systems Journal. © 1969 by International Business Machines Corporation.

tions. In particular, it may be invoked in a FORTRAN program (compiled on System/360 or 370) by the statement

CALL RANDOM(INT,REAL)

where INT is any fullword integer variable and REAL is any fullword variable (single precision). The integer variable, INT, should be given an initial value before the first use of the generator. The generator returns an integer random number in INT and a real random number between 0 and 1 in REAL. Tests indicate that the generator took about 31 microseconds per call on a System/360 model 67.

6-9

Generating Discrete Distributions

We demonstrate first how random numbers are used to generate sequences of numbers from a discrete distribution. When the discrete distribution is uniform, the requirement is to pick one of N alternatives with equal probability given to each. Given a random number $U(0 \leq U < 1)$, the process of multiplying by N and taking the integral portion of the product, which is denoted mathematically by the expression $[UN]$, gives N different outputs. The outputs are the numbers $0, 1, 2, \ldots, (N - 1)$. The result can be changed to the range of values C to $N + C - 1$ by adding C.

Alternatively, the next highest integer of the product UN can be taken. In that case, the outputs are $1, 2, 3, \ldots, N$. Note that the rounded value of the product $U \cdot N$ is *not* satisfactory. It produces $N + 1$ numbers as output, since it includes 0 and N, and these two numbers have only half the probability of occurring as the intermediate numbers.

Generally, the requirement is for a discrete distribution that is not uniform, so that a different probability is associated with each output. Suppose, for example, it is necessary to generate a random variable representing the number of items bought by a shopper at a store, where the probability function is the discrete distribution given previously in Table 6-1. A table is formed to list the number of items, x, and the cumulative probability, y, as shown in Table 6-5.

Taking the output of a uniform random number generator, U, the value is compared with the values of y. If the value falls in an interval $y_i < U \leq y_{i+1}$ $(i = 0, 1, \ldots, 4)$, the corresponding value of x_{i+1} is taken as the desired output. If the five values of uniformly distributed numbers derived from Table 6-3 in Sec. 6-6 were used, they would lead to the five outputs, 2, 2, 1, 5, 2.

It is not necessary that the intervals be in any particular order. A computer routine will usually search the table from the first entry. The amount of searching can be minimized by selecting the intervals in decreasing order of probability. Arranged as a table for a computer routine, the data of Table 6-5 would then appear

TABLE 6-5 Generating a Non-Uniform Discrete Distribution

No. of Items x	Probability $p(x)$	Cumulative Probability y
0	0	0
1	0.10	0.10
2	0.51	0.61
3	0.19	0.80
4	0.15	0.95
5	0.05	1.00

as shown in Table 6-6. With this arrangement, 51% of the searches will only need to go to the first entry, 70% to the first or second, and so on. With the original ordering, only 10% are satisfied with the first entry, and only 61% with the first two.

TABLE 6-6 Non-Uniform Discrete Distribution Data Arranged for a Computer Program

Cumulative Probability U	No. of Items x
0.51	2
0.70	3
0.85	4
0.95	1
1.00	5

Note that, since we are now discussing random number generators, rather than distribution functions, the input of Table 6-6 is being denoted by U to indicate a random number between 0 and 1. The output continues to be denoted by x, however, to maintain the connection to the distribution from which the numbers defining the generator have been derived.

6-10

Non-Uniform Continuously Distributed
Random Numbers

The general requirement in simulation is for a sequence of random numbers drawn from a distribution that is continuous and non-uniform. Most methods for deriving such numbers are based on the use of sequences of uniformly distributed random numbers. The most widely used method is based on an operation in statistics known as the probability integral transformation. The method, which is generally called the *inverse transformation method*, will be described in this section. A second general method, called the rejection method, is described in the next section. There are also various transformation methods, other than the one based on the probability integral, some of which will be discussed in the next chapter.

The probability integral transformation theorem can be stated as follows: If $U_i (i = 1, 2, \ldots)$ are independent, random variables, uniformly distributed over the interval 0 to 1, and $F^{-1}(x)$ is the inverse of the cumulative distribution function for the random variable X, then the random variables defined by $X_i = F^{-1}(U_i)$ are a random sample of the variable X; that is, to produce random numbers with a given distribution, the inverse cumulative distribution function must be evaluated with a sequence of uniformly distributed numbers in the range 0 to 1.

To demonstrate the validity of this result, consider a small interval Δx on the x axis between x and $x + \Delta x$, and the interval ΔF defined by the corresponding points $F(x)$ and $F(x + \Delta x)$, as illustrated in Fig. 6-7. For a number of the output sequence to fall in the interval Δx, the input number must fall in the interval ΔF.

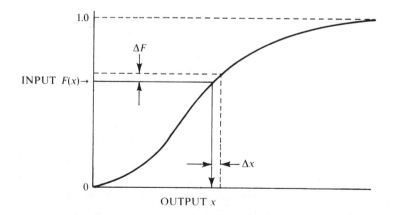

Figure 6-7. Generation of non-uniform continuous random numbers.

Since the input numbers are uniformly distributed over the range 0 to 1, the probability of an input number falling in the interval ΔF is equal to ΔF. The average value of the probability density function for x over the interval Δx is therefore $\Delta F/\Delta x$. In the limit as Δx tends to zero the value of the density function of x is dF/dx which by definition of $F(x)$ is the function $f(x)$.

If the density function $f(x)$ can be described mathematically, it is often possible to find an expression for the inverse of the cumulative distribution function which can then be evaluated with a sequence of uniformly distributed random numbers. Suppose, for example, the desired probability density function is

$$f(x) = a(1 - x) \qquad (0 \le x \le 1)$$
$$= 0 \qquad\qquad \text{elsewhere}$$

Then,

$$F(x) = a\left(x - \frac{x^2}{2}\right)$$

The value of a must be 2 in order to satisfy the condition that $\int_{-\infty}^{\infty} f(x)\, dx = 1$. Letting $U = F(x)$ and solving for x, we find the inverse function:

$$x = 1 - (1 - U)^{1/2}$$

Taking the 5 random numbers derived from Table 6-3 in Sec. 6-6, it will be found that the corresponding outputs are

U	x
0.1009	69.23
0.3754	88.48
0.0842	66.65
0.9901	142.33
0.1280	71.82

The operation of inversing an integral, however, cannot often be carried out to give a result that is easily evaluated from known mathematical functions, or from readily available tables of values. In the case of the distribution discussed in Sec. 6-3 and illustrated in Figs. 6-2 and 6-3, the cumulative distribution, although it can be expressed in mathematical form, cannot be easily inversed because it consists of a mixture of algebraic and exponential terms.

In that case, numerical computation methods must be used. To illustrate the method, consider the data for the length of telephone calls that were given in Table 6-2. The cumulative distribution of the lengths was given in Fig. 6-5 as a graph consisting of straight line segments. To generate random numbers representing telephone call lengths, this distribution must be inversed. The inversed graph, as shown in Fig. 6-8, appears in the conventional form, where the input, which is to be a uniformly distributed random number between 0 and 1, is shown as the abscissa and the output is the ordinate.

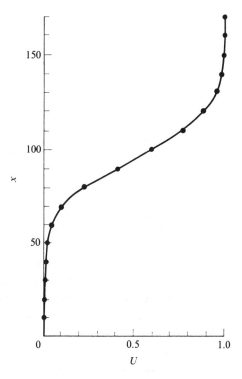

Figure 6-8. Inversed cumulative distribution of telephone call lengths.

Graphically, the process of generating the random numbers consists of taking as inputs a series of uniformly distributed random numbers, U_i, and reading the outputs, x_i, from the graph. As a computer program, this process must be translated into a table search procedure. The table input entries are the cumulative distribution values (column 5 of Table 6-2), and the output entries are the call length values at which the data were tabulated (column 1 of Table 6-2). Table 6-7 shows the data for telephone call length organized in this way. The first column gives the values of the cumulative distribution, and is labelled U, since a random number

TABLE 6-7   Generation of Telephone Call Lengths

U	x_t	a_t
0	0	10,000.00
0.001	10	10,000.00
0.002	20	10,000.00
0.003	30	5,000.00
0.005	40	1,250.00
0.013	50	357.14
0.041	60	153.85
0.106	70	82.64
0.227	80	57.14
0.402	90	50.76
0.599	100	57.14
0.774	110	82.64
0.895	120	153.85
0.960	130	357.14
0.988	140	1,111.11
0.997	150	5,000.00
0.999	160	10,000.00
1.000	170	—

between 0 and 1 is to supply the input. The second column is the output, x, representing the generated call length. The third column will be explained shortly.

Table 6-6 is similar to Table 6-7, but the former table was for generating discrete data. As was described in Sec. 6-9, if the input to that table fell between two tabulated values, the output was taken to be the value at the upper end of the interval. In the case of continuous data represented in tabular form, however, since all values between the tabulated values are possible, an interpolation between the tabulated values is made. The process is illustrated in Fig. 6-9. If the input entered in the table search falls in an interval between two tabulated input values, the output is taken to be the output value at the lower tabulated input plus an increment that divides the output interval in the same proportion that the input divides the input interval. This is consistent with the assumption that the cumulative distribution is approximated by straight line segments between the tabulated points. To carry out the process of interpolation numerically, it is necessary to know the slopes of these lines. Since these do not change once the function has been defined, they can be calculated and entered in the table used for generating the random numbers.

The cumulative distribution used in Table 6-7 gave a probability, y, as a function of call length, x. Because of the inversion of the distribution, the slope being sought is the slope of x with respect to y. The original function is defined at the points $x_i(i = 0, 1, \ldots, N)$ to have the values $y_i(i = 0, 1, \ldots, N)$. The value y_0 is assigned to have the value 0, and the last defined value y_N is assigned to be 1.

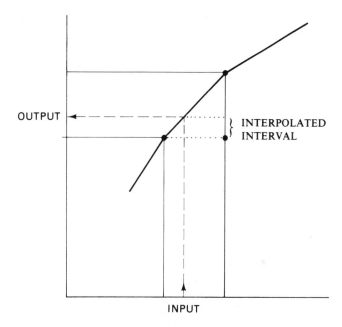

OUTPUT

INTERPOLATED
INTERVAL

INPUT

Figure 6-9. Illustration of interpolation.

The slopes, denoted by a_i, are defined as follows:

$$a_i = \frac{x_{i+1} - x_i}{y_{i+1} - y_i} \qquad (i = 0, 1, \ldots, N-1)$$

The values are entered in Table 6-7 as column 3. They represent the slopes per unit input, and they are stored on the line at which the slope begins. For example, the line that has the value 0.106 as input shows an output of 70 and a slope of 82.64. The slope of 82.64 is in effect over the input range from 0.106 to 0.227. Suppose the random number input to the table is 0.127. The output is then the output value at 0.106, which is 70, plus the interpolated increment of 82.64(0.127 − 0.106), which is equal to 1.75. The output is therefore 71.75 or, if the output is to be to the nearest second, 72. To express this more formally, the input, U, is matched against the values y_i that define the input. If the input should happen to equal one of the tabulated values, the corresponding output value, x_i, is taken. If U falls in the interval

$$y_i < U < y_{i+1} \qquad (i = 0, 1, \ldots, N-1)$$

then the output is

$$x = x_i + a_i(U - y_i)$$

Taking the 5 random numbers derived from Table 6-3, in Sec. 6-6, it will be found that the corresponding outputs are

U	x
0.1009	69.23
0.3754	88.48
0.0842	66.65
0.9901	142.33
0.1280	71.82

In collecting the original data on telephone call lengths, it was convenient to record the data to the nearest 10 seconds. When the data are inversed, these are not very convenient points at which to tabulate the results, as can be seen from Fig. 6-8, which shows the points crowded together at the beginning and at the end. Ideally, the points at which the function is defined should be selected so that the error introduced by the linear approximation is held within bounds while not using more points than are necessary for a given error level. If an approximation is being made to a function that is mathematically defined, it may be possible to carry out this optimization formally. With experimental data this is not possible. Some balance can be introduced, however, by merging intervals in which the cumulative distribution is only changing slowly. In the case of the telephone data, for example, some of the intervals were too small to plot in the bar histogram of Fig. 6-4. Rather than use the full data of Table 6-7, it would be more practical to use the fewer number of intervals shown in Fig. 6-4. A computer program for fitting points to an inverse cumulative distribution is given in Ref. (8).

6-11

The Rejection Method

Another method of generating random numbers, the *rejection method*, is applicable when the probability density function, $f(x)$, has a lower and upper limit to its range, a and b, respectively, and an upper bound c. The method can then be specified as follows:

a) Compute the values of two, independent uniformly distributed variates U_1 and U_2

b) Compute $X_0 = a + U_1(b - a)$

c) Compute $Y_0 = cU_2$

d) If $Y_0 \leq f(X_0)$, accept X_0 as the desired output; otherwise repeat the process with two new uniform variates.

The method is closely related to the process of evaluating an integral using the Monte Carlo technique, which was described in Sec. 3-2. Referring back to Fig. 3-1, it will be seen that the function to be integrated has the limits and the bound required for the rejection method. Like the function to be integrated, the probability density function is enclosed in a rectangle with sides of lengths $b - a$ and c. The Monte Carlo method required generating a point at random within the rectangle and deciding where the point lies with respect to the curve. The first three steps of the rejection method create just such a random point, and the last step relates the point to the curve of the probability density function. If the point falls on or below the curve, the value X_0 is accepted as being a sample from the desired distribution. Otherwise, the point is rejected and the process is repeated.

In the Monte Carlo method, acceptance means that the point is added to a count, n, and, after many trials, the ratio of that count to the total number of trials, N, is taken as an estimate of the ratio of the area under the curve to the area of the rectangle. In the rejection method, the curve is a probability density function, so the area under the curve must be 1. This means that the scale of the graph must be chosen so that the area of the rectangle is also 1; that is,

$$c(b - a) = 1$$

To demonstrate the validity of the method, Fig. 6-10 shows a probability density function with limits a and b, and upper bound c.[8] Given X_0, the method requires that X_0 be accepted as an output if $Y_0 \leq f(X_0)$. Consider a small interval of the x

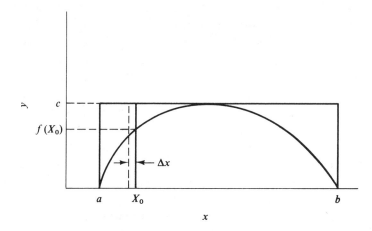

Figure 6-10. The rejection method.

[8]As in the case of the Monte Carlo method, it is not required that $f(x)$ take the value of the bound, c. For simplicity, we assume it is taken.

axis Δx at the point X_0. The acceptance condition can be expanded to mean that if Y_0 falls within the small rectangle formed by Δx, under the curve, then an output in the range $X_0 - \Delta x$ to X_0 will occur. If the range a to X_0 is covered by adjoining small intervals of this nature and the intervals are summed, it can be seen that the condition can be interpreted as meaning that if Y_0 falls on or below the curve between 0 and X_0, then outputs occur in the range a to X_0. In other words, the probability that the output X is less than or equal to X_0 is equal to the probability that Y_0 falls under the curve to the left of X_0, given that $X \leq X_0$.

The probability of X being less than or equal to X_0 is, by definition, $F(X_0)$. The probability of Y_0 falling on or below the curve to the left of X_0 is the ratio of the area under that part of the curve to the rectangle with sides $X_0 - a$ and c. Finally, since X_0 is uniformly distributed between a and b, the probability that X will be in the range a to X_0 is $(X_0 - a)/(b - a)$. We have, therefore,

$$F(X_0) = \frac{\int_a^{X_0} f(x)\, dx}{c(X_0 - a)} \cdot \frac{(X_0 - a)}{(b - a)}$$

Since $c(b - a) = 1$, it follows that

$$F(X_0) = \int_a^{X_0} f(x)\, dx$$

which demonstrates that X_0 has the desired distribution.

The inverse transformation method of generating random numbers requires that the desired probability density function be integrated and then inversed. The comment was made previously that it is not often that these operations result in a tractable function. The rejection method, however, operates only with the probability density function. If this is a known mathematical function, the rejection method is a convenient method of generating numbers without the need to carry out the approximations involved in setting up tables, such as Table 6-7, when the cumulative distribution cannot be inversed.

The rejection method, however, has the disadvantage that two uniform variates must be calculated for each trial point, and, since, some points are rejected, more than two uniform variates are needed for the creation of each output point. If the probability density function is not given by a mathematical function, but must be approximated numerically, the inverse transformation method is preferable, since it gives an output for each uniform variate that is generated.

Note that the correct application of the rejection method requires that the probability density function be limited. This means that the function is identically zero below a, and above b. Probability density functions that have long "tails," often stretching to infinity, are sometimes truncated, meaning that an arbitrary limit is made, and the function is set to zero beyond the limit. The truncated function is not strictly a probability density function, since it does not meet the condition that the area under the curve be 1. The error introduced is usually small, and can be ignored.

Exercises

In the following problems that require the use of a uniformly distributed random number generator, use the numbers given in Table 6-3, starting from the top of the leftmost column (column 1) and reading down the column.

6-1 Give the correct value of the constant A that makes the following equations for y a probability density function. Derive formulas for generating random numbers having these distributions and compute the first 10 values.

(a) $\quad y = \dfrac{1}{(x + A)}$ $\qquad\qquad 0 \le x \le 1$

$\qquad = 0$ $\qquad\qquad$ elsewhere

(b) $\quad y = 0.5 + A(x + 1.5)$ $\qquad 1 \le x \le 2$

$\qquad = 0$ $\qquad\qquad$ elsewhere

(c) $\quad y = A \sin x$ $\qquad\qquad 0 \le x \le \pi/2$

$\qquad = 0$ $\qquad\qquad$ elsewhere

(d) $\quad y = 0.25 + A(x - 1)$ $\qquad 1 \le x \le 2$

$\qquad = 0.25 - A(x - 3)$ $\qquad 2 < x \le 3$

$\qquad = 0$ $\qquad\qquad$ elsewhere

6-2 Approximate the following function with 10 straight lines at equally spaced intervals of x:

$$y = 0.5093 + 0.2 \sin x \qquad 0 \le x \le \pi/2$$

Use the approximation to derive 10 random numbers having this distribution.

6-3 Assuming a normal 365-day year, construct a table from which to generate a number between 1 and 12, inclusive, to represent the month of the year. Assume each day of the year is equally likely to be chosen, and number the months in their normal calendar sequence. Compute the first 10 outputs.

6-4 The probability of a batter swinging at a ball is 0.7. When he swings, the probability of his hitting is 0.6. If he hits the ball, the probability of its being caught is 0.5. Using columns 1, 2, and 3 of Table 6-3 to compute, respectively, the occurrence of swinging, hitting, and being caught, determine how many of the first 10 plays will result in a batter being out.

6-5 A uniformly distributed random integer between 1 and 20, inclusive, is to be generated from each of the numbers in column 1 of Table 6-3. How many of the outputs are prime numbers? (Include 1 as a prime number.)

6-6 Draw a relative frequency distribution from the following data, grouping the data in steps of 200. Find the mean and standard deviation of the data.

854	1128	411	194	2054
174	1268	1105	416	545
565	37	597	224	1180
870	245	559	133	99
416	382	421	3229	943
717	415	662	705	1498
1577	1714	521	484	324
1529	2863	16	151	1967
3073	273	3157	1354	288
1217	93	516	507	723

Bibliography

1 BROWN, G. W., *History of RAND's Random Digits*, NBS Applied Mathematics Series, no. 12, Washington, D. C.: National Bureau of Standards, 1951.

2 COVEYOU, R. R., and R. D. MACPHERSON, "Fourier Analysis of Uniform Random Number Generators," *J. ACM*, XIV, no. 1 (1967), 100–119.

3 CROXTON, FREDERICK E., DUDLEY J. COWDEN, and S. KLEIN, *Applied General Statistics*, (3rd. ed.), Englewood Cliffs, N. J.: Prentice-Hall, Inc., 1967.

4 DIANANDA, P. H., "Some Probability Limit Theorems With Statistical Applications," *Proc. Cambridge Philos. Soc.*, XLIX, (1953), 239–246. See footnote 4, this chapter.

5 GREENBERGER, MARTIN, "Methods in Randomness," *Commun. ACM*, VIII, no. 3 (1965), 177–179.

6 HULL, T. E., and A. R. DOBELL, "Mixed Congruential Random Number Generators for Binary Machines," *J. ACM*, XI, no. 1 (1964), 31–40.

7 HUTCHINSON, DAVID W., "A New Pseudorandom Number Generator," *Commun. ACM*, IX, no. 6 (1966), 432–433.

8 KISKO, THOMAS M., "An Automated Method of Creating Piecewise Linear Cumulative Probability Distributions," *Proc. Winter Simulation Conference, Dec. 6–8. 1976*, pp. 487–494. WSC/SIGSIM, New York Technical College, Farmingdale, N. Y.

9 LEHMER, D. H., "Mathematical Methods in Large-Scale Computing Units," *Proc. 2nd Annual Symposium on Large-Scale Digital Computing Machinery*, pp. 141–145, Cambridge, Mass.: Harvard University Press, 1951.

10 LEWIS, P. A. W., A. S. GOODMAN, and J. M. MILLER, "A Pseudo-Random Number Generator for the System/360," *IBM Syst. J.*, VIII, no. 2 (1969), 136–146.

11 NANCE, RICHARD E., and CLAUDE OVERSTREET, Jr., "Bibliography 29. A Bibliography on Random Number Generation," *Comput. Rev.*, XIII, no. 10 (1972), 495–508.

12 RAND Corporation. *A Million Random Digits With 100,000 Normal Deviates*, Glencoe, Ill.: The Free Press, 1955.

7

ARRIVAL PATTERNS AND SERVICE TIMES

7-1

Congestion in Systems

Most systems of interest in a simulation study contain processes in which there is a demand for service that causes congestion. There may, for example, be customers trying to check-out at a supermarket counter, work-pieces waiting for a machine to become available, ships waiting for a berth, and so on. The system can service entities at a rate which, in general, is greater than the rate at which entities arrive, but there are random fluctuations either in the rate of arrival, the rate of service, or both. As a result, there are times when more entities arrive than can be served at one time, and some entities must wait for service. The entities are then said to join a *waiting line*. The combination of all entities in the system—those being served, and those waiting for service—will be called a *queue*.

Congestion may be described in terms of three main characteristics. These are

(a) The *arrival pattern*, which describes the statistical properties of the arrivals.
(b) The *service process*, which describes how the entities are served.
(c) The *queuing discipline*, which describes how the next entity to be served is selected.

The service process, in turn, is described by two main factors: the *service time* and the *capacity*. The service time is the time required to serve an individual entity.

The service capacity, or, more simply, the capacity, is the number of entities that can be served simultaneously.

A third factor that may need to be described in discussing the service is its *availability*. The service may not be available at all times. For example, a machine may break down or be periodically removed from service for inspection. In that case, the availability will be a function of the system conditions. The availability may also be an intrinsic property of the service, as occurs, for example, with an elevator which is only able to admit people when it is stopped with its doors open.

To model a system, the probability functions that describe the arrival patterns and the service times must be given. Most system models consist of several activities, interconnected by having the output of one become the input for another. The arrival patterns that result from the transfer of entities between these activities arise from endogenous events and do not have to be described; they are generated as the simulation proceeds. The exogenous arrivals coming to the system from its environment, however, do need to be described.

When measurements have been taken of an arrival pattern, the approximation methods described in Sec. 6-10 can be used to reproduce the distribution. This is likely to be done when studying a specific system. When studying general types of systems, it is desirable to have some fundamental representation of the distribution which, while it may be theoretical, has the general characteristics of the traffic occurring in that type of system. A number of theoretical distributions that serve this purpose will be discussed. In addition, the introduction of stochastic variables into a model makes most of the system performance measures vary stochastically. The measurement of queues is among the most important outputs of a simulation, and the ways of describing and measuring queues will also be discussed.

In simple cases, the statistical properties of the arrivals and the service are independent of time, in which case they are said to be *stationary*. There may, however, be effects that cause these factors to vary with time, in which case the process is said to be nonstationary or *time-variant*. For example, the arrival rate may depend on time of day, resulting in peak load periods, or the rate of service may speed up when there are long queues. Where this happens, the effect can be simulated by making some parameter of the distribution, such as the mean, vary according to conditions.

7-2
Arrival Patterns

The usual way of describing an arrival pattern is in terms of the *inter-arrival time*, defined as the interval between successive arrivals. For an arrival pattern that has no variability, the inter-arrival time is, of course, a constant. Where the arrivals vary stochastically, it is necessary to define the probability function of the inter-arrival times. Two or more arrivals may be simultaneous. If n arrivals are simultaneous, then $n - 1$ of them have zero inter-arrival times.

In discussing arrival patterns, the following notation will be used:

T_a Mean inter-arrival time
λ Mean arrival rate

They are related by the equation

$$\lambda = \frac{1}{T_a}$$

For example, records might show that an office, working an eight-hour day for five days a week, gets about 800 telephone calls a week. Suppose a model of the office during working hours is to be constructed, using a time scale of minutes. A week has 40 working hours, so 800 calls implies an average inter-arrival time of $40 \times 60/800 = 3$ minutes; which is equivalent to an arrival rate of $\frac{1}{3} = 0.333$ calls per minute.

Probability distributions have been described so far as either probability density functions or cumulative distribution functions. When describing arrival patterns, it is common practice to express the distribution in terms of the probability that an inter-arrival time is greater than a given time. We define $A_0(t)$ as the *arrival distribution*, so that:

$A_0(t)$ is the probability that an inter-arrival time is greater than t.

Since the cumulative distribution function $F(t)$ is the probability that an inter-arrival time is less than t, it is related to the arrival distribution by

$$A_0(t) = 1 - F(t)$$

From its definition, the function $A_0(t)$ takes a maximum value of 1 at $t = 0$ and it cannot increase as t increases.

7-3

Poisson Arrival Patterns

A common situation is that the arrivals are said to be completely random. Speaking loosely, this means that an arrival can occur at any time, subject only to the restriction that the mean arrival rate be some given value. More formally, it is assumed that the time of the next arrival is independent of the previous arrival, and that the probability of an arrival in an interval Δt is proportional to Δt. If, in fact, λ is the mean number of arrivals per unit time, then the probability of an arrival in Δt is $\lambda \Delta t$. With these assumptions, it is possible to show that the distribution of the inter-arrival times is exponential. The probability density function of the inter-arrival time is given by

$$f(t) = \lambda e^{-\lambda t} \qquad (t \geq 0) \tag{7-1}$$

It follows that the arrival distribution is

$$A_0(t) = e^{-\lambda t} \tag{7-2}$$

The number λ is the mean number of arrivals per unit time. The actual number of arrivals in a period of time t is a random variable. It can be shown that with an exponential distribution of inter-arrival times, the probability of n arrivals occurring in a period of length t is given by

$$p(n) = \frac{(\lambda t)^n e^{-\lambda t}}{n!} \qquad (n = 0, 1, 2, \ldots) \tag{7-3}$$

This distribution, which is called the *Poisson distribution*, is discrete. The exponential distribution is, of course, continuous, since the inter-arrival time can take any non-negative value. Because of this connection between the two distributions, a random arrival pattern is often called a *Poisson arrival pattern*. Where this term is used, it will mean that the inter-arrival time is exponentially distributed.

As an illustration, Table 7-1 lists the times at which 20 successive customers

TABLE 7-1 Arrival Data

Arrival Number	Arrival Time	Inter-arrival Time	Arrival Number	Arrival Time	Inter-arrival Time
1	12.5	12.5	11	136.4	21.4
2	15.1	2.6	12	142.7	6.3
3	44.1	29.0	13	151.2	8.5
4	62.6	18.5	14	162.5	11.3
5	65.3	2.7	15	167.2	4.7
6	67.6	2.3	16	172.9	5.7
7	71.0	3.4	17	179.8	6.9
8	92.5	21.5	18	181.6	1.8
9	106.5	14.0	19	185.0	3.4
10	115.0	8.5	20	194.9	9.9

arrive at a store. Column 1 lists the customer number, column 2 gives the time of arrival to the nearest tenth of a minute. The third column gives the inter-arrival times between the customers, the inter-arrival time for the first customer being measured from time zero.

In Sec. 5-2 we discussed exponential growth models and described a method for testing whether data are exponentially distributed. The method consists of ranking the data by magnitude and plotting it on semi-logarithmic paper. We use the same method here. The inter-arrival times range from a low of 1.8 to a high of 29.0. Plotting the values against rank on semi-logarithmic paper gives the results shown in Fig. 7-1. The data fall close to the straight line shown in the figure (which has been drawn freehand.) Accepting that the data are exponentially distributed, the mean inter-arrival time, T_a, is estimated by taking the average of the observed values to get a value of 9.74. The estimated arrival rate, therefore, is $\lambda = 1/T_a = 0.103$.

To demonstrate the Poisson distribution of the arrivals, consider the interval 0 to 200 minutes broken into adjoining 20 minute periods, and count the number of arrivals in each successive period. (It is assumed that the next arrival, after number 20, occurs after time 200.) In Table 7-2, we collect together the results arranged in time period order. Table 7-3 rearranges the data of Table 7-2. Column 1 gives the number of arrivals, ranging from 0 to 5. The second column shows how many periods had that many arrivals.

TABLE 7-2   Number of Arrivals in Successive Periods

Time Period	Number of Arrivals	Time Period	Number of Arrivals
0 — 20	2	100 — 120	2
20 — 40	0	120 — 140	1
40 — 60	1	140 — 160	2
60 — 80	4	160 — 180	4
80 — 100	1	180 — 200	3

Referring now to Eq. (7-3), which defines the Poisson distribution: for the present case, $\lambda = 0.103$, and $t = 20$. Computing the value of $p(n)$ for $n = 0, 1, 2, 3, 4$, and 5[1] gives the values shown in column 3 of Table 7-3. These figures are the probabilities of getting the given number of arrivals in a 20 minute period when the arrival rate is 0.103 per minute. The particular sample used in this case was for 10 intervals, so, multiplying column 3 by 10, gives the expected number of intervals,

[1]The function $p(n)$ has a value for all non-zero values of n, but it becomes very small beyond $n = 5$, in the present case.

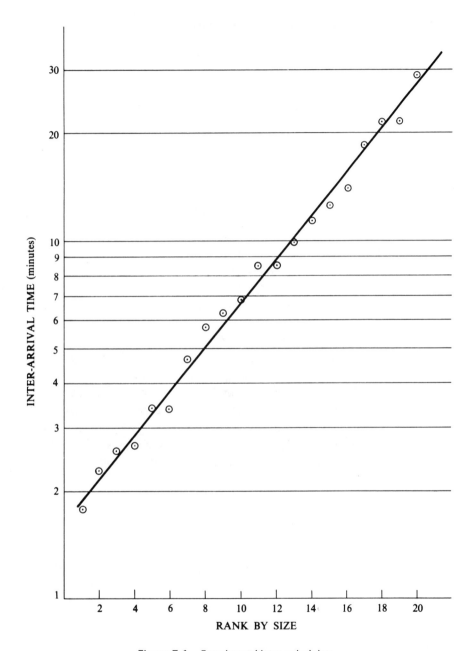

Figure 7-1. Experimental inter-arrival data.

TABLE 7-3 Poisson Distribution Data

Number of Arrivals in 20 Mins.	Actual Number of Occurrences	$p(n)$	Expected Number of Occurrences
0	1	0.128	1.3
1	3	0.266	2.7
2	3	0.272	2.7
3	1	0.187	1.9
4	2	0.096	1.0
5	0	0.040	0.4

shown in column 4. Comparing columns two and four, it can be seen that the actual values are close to the values that are to be expected.

The Poisson arrival pattern has great practical importance because it is found to represent arrivals in many types of systems. It also has importance in the theory of queues, because the underlying assumption that an inter-arrival time is independent of previous arrivals enables mathematical solutions to be obtained in a number of cases.

The Poisson distribution is not only of importance in discussing arrival patterns, it is one of the most widely used distributions in the application of probability theory and has been used to describe many different phenomena, (5). In particular, although the Poisson distribution has been described here as a representation of a random arrival pattern, it is also used to describe a service pattern. When a service time is considered to be completely random, it may be represented by an exponential distribution. In that case, the associated Poisson distribution represents the number of entities served within a given interval.

7-4

The Exponential Distribution

The cumulative distribution function of the exponential distribution is given by

$$y = 1 - e^{-\lambda t}$$

which can be inversed to give

$$\lambda t = -\ln(1 - y)$$

where the symbol $\ln$ denotes the natural logarithm of a number; that is, the logarithm to the base e. Because y represents a cumulative distribution, the term

$1 - y$ has values between 0 and 1. Over that range of values, a logarithm to any base is negative. The negative sign in the formula therefore results in a positive number. If the more commonly used logarithms to the base 10 are used, the symbol *log* is used, and the formula takes the form[2]

$$\lambda t = -0.4343 \log (1 - y)$$

However, when using a digital computer, it is as easy to generate natural logarithms as to generate logarithms to the base 10, so the natural logarithm form will be used here.

Using the natural logarithm formula and substituting for y a series of uniformly distributed random numbers between 0 and 1 gives as output a series of random numbers that are exponentially distributed. If the numbers y are uniformly distributed, so also are the numbers $1 - y$, so it is possible to use the simpler formula

$$t = \frac{-ln(y)}{\lambda} = -T_a \, ln(y)$$

It will be noticed that the exponential distribution is completely characterized by one parameter, its mean value T_a, that appears as a multiplier in the formula for generating exponentially distributed random numbers. Numbers with any mean can be derived from a generator of mean value 1 by multiplying its output by the required mean. This is a specific property of the exponential distribution and cannot generally be applied to other distributions.

Denoting the required mean value by AVR, and using the uniform random number generator described in Sec. 6-8, a suitable FORTRAN subroutine is

```
CALL RANDOM(INT.REAL)
X = -AVR*ALOG(REAL)
```

The expression **ALOG(REAL)** represents a function supplied with FORTRAN for the (natural) logarithm of a real number. The result X will appear as a floating-point number, unless it has been declared otherwise.

The appearance of this formula for computing exponentially distributed numbers is deceptively simple. The number of programming statements is small, but the time to execute them can be relatively long. The logarithm function can require a large number of instruction executions, because it is determined from a

---

[2]If logarithms to the base 10 are used, it must be remembered that conventional logarithm tables are for numbers in the range 1 to 10. An exponent transformation is needed outside that range. For example, $\log 5 = 0.6990$, therefore $\log 0.5 = 0.6990 - 1 = -0.3010$.

series that converges slowly. A simulation may need thousands or even hundreds of thousands of random numbers for its completion, so there is a premium on keeping the time required to evaluate a number as small as possible. As a result, many methods have been devised for reducing the time by substituting, either wholly or in part, table searches of the type discussed in Sec. 6-10. Naturally, these methods require more storage space in a computer and result in some loss of accuracy; but, where the slight loss of accuracy can be accepted, they can lead to substantial time savings.

As an example, Table 7-4 shows a simple table of values for the function $ln(1 - y)$, (7). Since the table has to be constructed beforehand, there is no advantage in using the alternative form $ln y$, as there would be if the function were being evaluated for each number. The accuracy obtained after multiplying by the required mean value T_a is about 0.1% when $T_a > 250$ and about 1% when $45 < T_a \leq 250$.

TABLE 7-4 Table for Generating the Exponential Distribution

Input	Output	Slope	Input	Output	Slope
0	0	1.04	0.90	2.30	11.0
0.1	0.104	1.18	0.92	2.52	14.5
0.2	0.222	1.33	0.94	2.81	18.0
0.3	0.355	1.54	0.95	2.99	21.0
0.4	0.509	1.81	0.96	3.20	30.0
0.5	0.690	2.25	0.97	3.50	40.0
0.6	0.915	2.85	0.98	3.90	70.0
0.7	1.20	3.60	0.99	4.60	140
0.75	1.38	4.40	0.995	5.30	300
0.80	1.60	5.75	0.998	6.20	800
0.84	1.83	7.25	0.999	7.0	3,333
0.88	2.12	9.00	0.9997	8.0	—

The approximation in Table 7-4 has truncated the exponential distribution so that the largest possible value is 8. The probability of a sample from an exponential distribution being more than eight times the mean value is less than 10^{-4}. For some other methods of generating exponential variates, see (1) and (10).

As an example, Table 7-5[3] shows how arrival times would be generated, assuming exponentially distributed inter-arrival times with a mean of 20 minutes, and using Table 7-4 to generate the distribution. Column 1 is a set of uniformly distributed random numbers. They are, in fact, the first 10 numbers of the first column of Table 6-3, truncated to the first three figures. The second column gives the output

[3]Table 7-5 taken from Ref. (6), Chap. 3.

TABLE 7-5 Calculation of Arrival Times

Uniformly Distributed Random Numbers	Exponentially Distributed Random Numbers (mean = 1)	Inter-Arrival Times (mean = 20)	Arrival Times (min.)
0.100	0.104	2.08	2.08
0.375	0.470	9.40	11.48
0.084	0.087	1.74	13.22
0.990	4.600	92.00	105.22
0.128	0.137	2.74	107.96
0.660	1.086	21.72	129.68
0.310	0.370	7.40	137.08
0.852	1.963	39.26	176.34
0.635	1.915	38.30	214.64
0.737	1.333	26.66	241.30

of Table 7-3, using the method described in Sec. 6-10 for generating nonuniformly distributed random numbers. The results are exponentially distributed numbers with a mean of 1. The third column multiplies these numbers by 20 to get exponentially distributed numbers with a mean of 20, and the last column accumulates the inter-arrival times of column 3 to get the actual arrival times.

7-5

The Coefficient of Variation

The coefficient of variation was introduced in Sec. 6-4, when discussing ways in which probability distributions are measured. Applied to arrival times, it is the ratio of the standard deviation of the inter-arrival time, s_a, to the mean inter-arrival time, that is s_a/T_a.[4] The coefficient of variation, it will be recalled, is a dimensionless measure of the degree to which data are dispersed about the mean. A coefficient of zero means there is no variation, that is, the data have a constant value. As the coefficient increases the data become more dispersed.

For the case of an exponential distribution of mean value T, the standard deviation is also T. The coefficient of variation, therefore, is 1. If the coefficient of variation of the measured data is found to be close to 1, it is reasonable to suppose that an exponential distribution may fit the data. A chi-square test should be carried out, however, to test the level of significance to which the data fits. Two other

[4]For convenience, we now drop the suffixes.

theoretical distributions that are often used when the coefficient of variation is significantly less than or greater than 1 are the Erlang and hyper-exponential distributions, respectively. These will be discussed in the next two sections.

7-6

The Erlang Distribution

There is a class of distribution functions named after A. K. Erlang, who found these distributions to be representative of certain types of telephone traffic. An arrival pattern governed by this type of function has the following density function and arrival distribution [(13), p. 41]

$$f(t) = (k\lambda)^k \left[\frac{e^{-k\lambda t}}{(k-1)!} \right] t^{k-1}$$

$$A_0(t) = e^{-k\lambda t} \sum_{n=0}^{k-1} \frac{(k\lambda t)^n}{n!}$$

(7-4)

where k is a positive integer greater than zero. Figure 7-2 illustrates the Erlang arrival distribution for several values of k. Putting $k = 1$ in Eq. (7-4) and comparing with Eq. (7-1), it can be seen that the case of $k = 1$ is the exponential distribution. The standard deviation of the Erlang distribution is $T/k^{1/2}$, so that the coefficient of variation is $1/k^{1/2}$. The coefficient has a maximum value of 1 for the case of the exponential distribution, and it decreases as k increases. It follows that

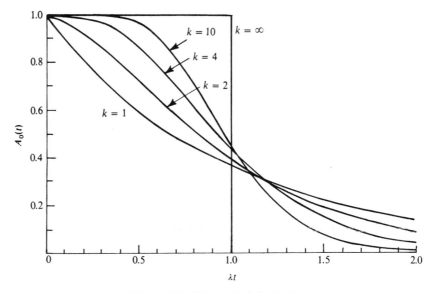

Figure 7-2. Erlang arrival distributions.

data represented by an Erlang distribution ($k > 1$) cluster more closely to the mean value than exponentially distributed data, so that there are fewer low values and high values. Ultimately, as k tends to infinity, the Erlang distribution tends to the case of a constant arrival time, for which the arrival distribution is the step function shown in Fig. 7-2.

When measured data are found to have a coefficient of variation significantly less than 1, it is reasonable to suppose that the data may be represented by the Erlang distribution for which k is closest to the value $(T/s)^2$. Again, the significance of the fit should be tested with a chi-square test.

While Eq. (7-4), describing the Erlang density function, may appear to be rather complex, the distribution can be given a simple physical interpretation, as illustrated in Fig. 7-3 [(13), p. 44]. Suppose there are k stages of service arranged in series, each having an exponentially distributed service time with the same mean value T/k. When an entity is given service, it passes through all k stages with a random and independently selected service time in each stage. The passage from one stage to the next occurs without loss of time, and the next entity is not allowed to enter the first stage until the preceding entity has cleared all stages. It can be shown that the distribution of overall service time is an Erlang distribution of the kth order with a mean service time T.[5] From this analogy, it is apparent that a number from the Erlang–k distribution with mean value T can be created by generating k independent exponentially distributed random numbers, each with mean value T/k, and summing the k numbers.

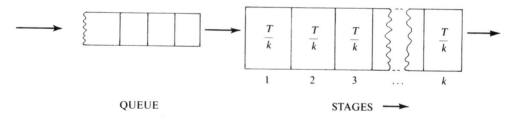

QUEUE   STAGES ⟶

Figure 7-3. Model for Erlang distributions.

7-7

The Hyper-Exponential Distribution

The hyper-exponential distribution defines a class of distributions that have a standard deviation greater than their mean value; consequently, they can be used to match data found to have a coefficient of variation greater than 1. They represent

[5]In some descriptions of the Erlang distribution, T is specified as the mean time of the individual exponential stages. The mean time of the overall process is then kT.

data where low and high values occur more frequently than with an exponential distribution; the data distribution may in fact be bimodal, showing a peak on either side of the mean value.

As with the Erlang distributions, it is possible to give a physical analogy to explain the nature of the distribution [(13), p. 52]. Suppose that there are two parallel stages of processing, as shown in Fig. 7-4. Both stages have exponential

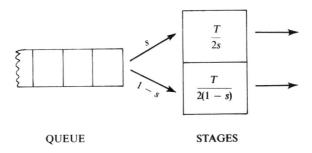

QUEUE   STAGES

Figure 7-4. Model for hyper-exponential distributions.

service times, one with a mean value of $T/2s$ and the other with mean value $T/2(1 - s)$, $0 < s \leq \frac{1}{2}$. When an entity is to be served, a random choice is made between the two stages; the probability of choosing the stage with mean of $T/2s$ being s and the probability of choosing the other being $(1 - s)$. A second entity cannot enter either stage until the preceding entity has been cleared. The equation for the resultant arrival distribution is found to be [(13), p. 53]

$$A_0(t) = se^{-2s\lambda t} + (1 - s)e^{-2(1-s)\lambda t} \qquad (0 < s \leq \tfrac{1}{2})$$

The variance is kT^2 so that the coefficient of variation is $k^{1/2}$, where

$$k = \frac{(1 - 2s + 2s^2)}{2s(1 - s)} \qquad (0 < s \leq \tfrac{1}{2})$$

When $s = \frac{1}{2}$, the value of k is 1 and the distribution becomes the exponential distribution. Figure 7-5 illustrates some hyper-exponential arrival distributions for various values of k. The corresponding values of s are also shown in Fig. 7-5.

Hyper-exponential distributions of random numbers can be generated by the process described in Fig. 7-4. An exponentially distributed random number with mean value of 1 is generated from a uniformly distributed random number. A second uniformly distributed random number is compared to s. If it is less than s, the exponentially distributed random number is multiplied by $T/2s$; otherwise, it is multiplied by $T/2(1 - s)$.

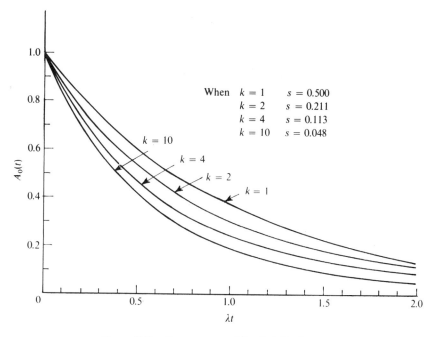

$$\text{When}\quad \begin{array}{ll} k = 1 & s = 0.500 \\ k = 2 & s = 0.211 \\ k = 4 & s = 0.113 \\ k = 10 & s = 0.048 \end{array}$$

Figure 7-5. Hyper-exponential arrival distributions.

7-8

Service Times

Frequently, the service time of a process is constant; but, where it varies stochastically, it must be described by a probability function. In discussing service times, the following notation will be used:

T_s = Mean service time

μ = Mean service rate

$S_0(t)$ = Probability that service time is $> t$

If the service time is considered to be completely random, it may be represented by an exponential distribution or, if the coefficient of variation is found to differ significantly from 1, an Erlang or hyper-exponential distribution may be used. A common situation is that, although the service time should be a constant, there are random fluctuations due to uncontrolled factors. For example, a machine tool may be expected to take a fixed time to turn out a part, but random variations in the amount of material to be removed and the toughness of the material cause fluctuations in the processing time. The normal, or Gaussian, distribution is often used to represent the service time under these circumstances.

7-9

The Normal Distribution

The *normal probability density function* is symmetric about its mean value and is completely characterized by its mean value μ and standard deviation σ. If the transformation $z = (x - \mu)/\sigma$ is made, the distribution is transformed to a form in which the mean is 0 and the standard deviation is 1. In this form, the probability density function is

$$f(z) = \frac{1}{(2\pi)^{1/2}} e^{-z^2/2}$$

It is customary to create generators that determine numbers distributed according to the function $f(z)$ and to derive the variable x that has mean μ and standard deviation σ by the transformation

$$x = z\sigma + \mu \tag{7-5}$$

The variable z is usually called the *standard normal variate*.

Neither the cumulative distribution function nor its inverse can be expressed in terms of simple mathematical functions, but approximation methods are available for the generation of standard normal variates. A very useful approximation method derives normally distributed random numbers by summing several uniformly distributed random numbers x_i according to the following formula (6):

$$y = \frac{\sum_{i=1}^{k} x_i - \frac{k}{2}}{\left(\frac{k}{12}\right)^{1/2}}$$

The distribution of a sequence of numbers derived from this formula approaches a normal distribution with mean zero and standard deviation 1 as k approaches infinity. Quite small values of k give good accuracy. A convenient value of k is $k = 12$, in which case, the formula takes the form

$$y = \sum_{i=1}^{12} x_i - 6.0$$

This is a direct method of deriving normally distributed random numbers, based upon the properties of random numbers rather than using the inverse transformation method. The following FORTRAN subroutine is based on this method and produces a number V which is normally distributed with mean AM and standard deviation S:

```
           SUBROUTINE GAUSS(S,AM,V)
           A = 0.0
           DO 50 I = 1,12
           CALL RANDOM(INT,REAL)
      50   A = A + REAL
           V = (A − 6.0)*S + AM
           RETURN
           END
```

The subroutine is used by issuing the statement

```
           CALL GAUSS(S,AM,V)
```

The subroutine RANDOM, called within GAUSS, is the uniform random number generator discussed in Sec. 6-8.

Another useful method, that is based on a transformation other than the inverse transformation method, is capable of giving normal variates to a high degree of accuracy, (2). Given two independent uniform variates, the method generates two outputs, as follows

$$X_1 = (-2lnU_1)^{1/2} \cos 2\pi U_2$$
$$X_2 = (-2lnU_1)^{1/2} \sin 2\pi U_2$$

The outputs are independent, and each is from a normal distribution with a mean of 0 and a standard deviation of 1. Other methods are given in (3) and (11).

If the inverse transformation method is used, Table 7-6 gives a suitable approximation to the standard normal variate. It truncates the values of z at ± 3.5. The probability of the normal variate falling outside this range is less than 10^{-4}.

Note that, whichever method is used for generating the standard normal variate, it is possible to get negative outputs from the transform defined by Eq. (7-5). If the mean is positive and z is negative, which happens 50% of the time, then Eq. (7-5) will give a negative output—if the mean is not more than $-z$ times the standard deviation. Using Table 7-6, the mean must be at least 3.5 times the standard deviation to avoid negative outputs. In the case of the subroutine GAUSS, the factor is six times.

It may be necessary to truncate the output x at some minimum value, in particular, at zero to avoid negative numbers. If so, the part of the normal distribution

TABLE 7-6 Table for Generating a Standard Normal Variate

Input	Output	Slope	Input	Output	Slope
0.0000	−3.50	384.60	0.5000	0.00	2.53
0.0013	−3.00	113.63	0.5987	0.25	2.69
0.0035	−2.75	92.38	0.6915	0.50	3.05
0.0062	−2.50	41.67	0.7734	0.75	3.68
0.0122	−2.25	23.58	0.8413	1.00	4.71
0.0228	−2.00	14.42	0.8944	1.25	6.44
0.0401	−1.75	9.36	0.9332	1.50	9.36
0.0668	−1.50	6.44	0.9599	1.75	14.42
0.1056	−1.25	4.71	0.9772	2.00	23.58
0.1587	−1.00	3.68	0.9878	2.25	41.67
0.2266	−0.75	3.05	0.9938	2.50	92.38
0.3085	−0.50	2.69	0.9965	2.75	113.63
0.4013	−0.25	2.53	0.9987	3.00	384.60
			1.000	3.50	—

that is accepted should be adjusted to allow for the fact that the area under the curve is no longer equal to 1, and, therefore, is no longer a proper probability density function. The truncations introduced in Table 7-5 and implied in the subroutine GAUSS should also be compensated for, but the magnitudes of the adjustments, compared with the intrinsic errors of the methods, are too small to be worth making.

An example where this type of adjustment is necessary occurs in the simulation of disk storage devices that move an arm over tracks to locate data. A certain minimum time is needed to start the arm moving. The time to reach the desired track is then proportional to the number of tracks that must be crossed. The time to locate the data can, therefore, be expressed as

$$t = a + bn \qquad (7\text{-}6)$$

where a and b are constants, and n is the number of tracks to be crossed. If the device is designed to start each search from the track it last read, and it is assumed that the requests for data are uniformly distributed over the range of tracks being searched, N, it can be shown that n is a random variable that is approximately normally distributed with a mean of $N/3$ and a standard deviation of $N/\sqrt{18}$, (12). It can also be shown that if n is normally distributed, so also is t, as defined by Eq. (7-6). The mean of t is a plus b times the mean of n, and the standard deviation of t is b times the standard deviation of n.

In the case of the IBM 2314 disk file, Fig. 7-6, which is adapted from (12), and uses data published in (8), shows that the seek time can be approximated by the equation

$$t = 45 + 0.43n$$

The IBM 2314 disk file has 200 tracks on a disk. Suppose, however, the search for data is confined to 100 adjacent tracks, and the data is uniformly spread over these tracks. The normal distribution from which t is drawn then has a mean of 58.3, and a standard deviation of 10.1. The value of t, however, must be truncated at 45. The value 45 is only 13.3 below the mean value, that is, only 1.32 standard deviations below the mean. A value of z less than -1.32 from any of the random number generators will cause a value of x that is less than 45. If x is truncated so that it is never less than 45, it is equivalent to dropping the part of the standard normal probability density function that falls below $z = -1.32$. Tables of the standard normal variate will show that the area under that part of the curve is .093. To make the output of the generator, after truncation, more representative of a true probability density function, the output should be multiplied by $1/(1 - 0.093) = 1.1$.

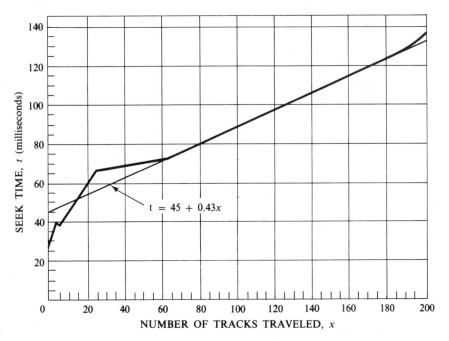

Figure 7-6. Time to move an IBM 2314 disk drive arm.

7-10

Queuing Disciplines

The third factor for describing congestion is the queuing discipline that determines how the next entity is selected from a waiting line. The most common queuing disciplines and some of the terms used in describing queues will now be discussed.

(a) A *First-In, First-Out* discipline or, as it is commonly abbreviated, FIFO, occurs when the arriving entities assemble in the time order in which they arrive. Service is offered next to the entity that has waited longest.

(b) A *Last-In, First-Out* discipline, usually abbreviated to LIFO, occurs when service is next offered to the entity that arrived most recently. This is approximately the discipline followed by passengers getting in and out of a crowded train or elevator. It is the precise discipline for records stored on a magnetic tape that are read back without rewinding the tape.

(c) A *Random* discipline means that a random choice is made between all waiting entities at the time service is to be offered. Unless specified otherwise, the term random implies that all waiting entities have an equal opportunity of being selected.

It has been implied that an entity that joins a waiting line stays until it is served. This may not happen; for example, a customer may become impatient and leave. The term *reneging* is used to mean that entities leave before their service is due to begin, and when this occurs the rules for reneging must be specified. Reneging may depend on the waiting line length or the amount of time an entity has waited. The term *balking* is used to mean an entity refuses to join a waiting line.

When there is more than one line forming for the same service, the action of sharing service between the lines is called *polling*. Examples of polling systems are a bus stopping along a route to pick up passengers, a clerk circulating around a group of offices to pick up mail, or a computer scanning a number of input terminals to detect the presence of input messages. The polling discipline needs to be specified by giving the order in which the lines are served, the number of entities served at each offering of service, and the time (if any) in transferring service between lines.

Some members of a waiting line may have priority, meaning that they have a right to be served ahead of all other members of lower priority. There may be several levels of priority and, since there may be several members in the same priority class, a queuing discipline must be specified for the members within a class. A typical situation is for service to be by priority, and first-in, first-out within priority class. The priority may be a preassigned attribute of the entity, as, for example, the rule, women and children first, or it may depend upon the waiting line. For example, priority may depend upon the length of time an entity has waited in a polling system.

When priority is allowed, a factor that must be considered is what happens when a new arrival has a higher priority than the entity currently being served. The priority may only control the position of the new arrival in the waiting line. In some cases, however, it may allow the new arrival to displace the entity being served. The new arrival is then said to *interrupt* or *preempt* the service. The term interrupt is usually interpreted as meaning that, when the interrupting entity has finished its service, the service is returned to the entity that was interrupted. Preemption usually means the preempted entity is displaced, either out of the system or back to the waiting line. However, these terms are not used consistently and their exact meaning should be determined when they are used.

7-11

Measures of Queues

We have already defined the mean inter-arrival time, T_a, and the mean service time, T_s, together with the corresponding rates, λ and μ. The ratio of the mean service time to the mean inter-arrival time is called the *traffic intensity*, denoted by u. As has just been mentioned, if there is any balking or reneging, not all arriving entities will get served. It is necessary, therefore, to distinguish between the actual arrival rate and the arrival rate of the entities that get served.

As the name might suggest, the term traffic intensity is used extensively when discussing telephone traffic. The unit of measurement for traffic intensity is, in fact, the *Erlang*, named after a pioneer in the study of telephone systems. Using the relationships between times and rates, traffic intensity can be denoted symbolically in various ways. The notation used most often in telephone studies is $\lambda' T_s$. Here, the prime is denoting all arrivals, including losses. Removing the prime to indicate the arrival rate of entities that stay to be served, defines a quantity called the *server utilization*. It is denoted by ρ, and most commonly, is expressed in terms of the rates of arrival and service. We, therefore, have the following definitions:

$$u = \lambda' T_s \qquad \text{Traffic intensity}$$

$$\rho = \lambda T_s = \frac{\lambda}{\mu} \qquad \text{Server utilization (single server)}$$

As defined, both the traffic intensity and the server utilization can be greater than 1. If so, a single server cannot keep up with the flow of traffic. One way of describing the figure for traffic intensity, in fact, is to say it is the minimum number of servers needed to handle the traffic with no losses. If more than one server is used, say n servers, the server utilization is redefined to reflect the load on each individual server, as follows

$$n \qquad \text{Number of servers}$$

$$\rho = \frac{\lambda}{n\mu} \qquad \text{Server utilization}$$

A server utilization of 1 or less then implies that the system can keep up with the traffic flow. However, as we shall see, values of the server utilization approaching 1 are undesirable because they are associated with long waiting lines.

Two principal measures of queuing systems are the mean number of entities waiting and the mean time they spend waiting. Both these quantities may refer to the total number of entities in the system, those waiting and those being served, or they may refer to the entities in the waiting line only. These quantities are denoted as follows:

L Mean number of entities in the system
L_w Mean number of entities in the waiting line
W Mean time to complete service (including the wait for service)
W_w Mean time spent waiting for service to begin

The probability that an entity will have to wait more than a given time, called the *delay distribution*, is also of interest and will be denoted as follows:

$P(t)$ Probability that the time to complete service is greater than t
$P_w(t)$ Probability that the time spent waiting for service to begin is greater than t.

7-12

Mathematical Solutions of Queuing Problems

A number of problems involving queues can be solved mathematically. It is beyond the scope of this text to discuss these in any detail. For further details, the reader is referred to (4), (9), and (13). Most of the solutions obtained are for a single activity servicing exogenous arrivals. Explicit solutions have been obtained in cases where the arrival pattern and the service times are constant, exponential, Erlangian, and hyper-exponential. Most activities in a simulation model receive their input from the output of other activities, and the conditions of the mathematical solutions are not applicable. Nevertheless, there are times when the conditions of the mathematical solutions are approximately true, and the known solutions provide some guidance on the performance to be expected. They will also provide a means of describing some of the considerations involved in designing service systems. All solutions given here are for infinite populations of sources; that is, the arrival rates are not affected by the number of entities currently in the system. They are also for lossless systems.

We first give solutions for the case of a single server with Poisson arrivals and different service time distributions. The cases considered are for hyper-exponential, exponential, Erlang, and constant service time distributions. (See Sec. 7-2, 7-6, 7-7, and 7-11 for notation.) The equations for the mean number of entities in the system, when $\rho < 1$, are

$$L = \frac{\rho(2 - \rho)}{2(1 - \rho)} \qquad \text{Constant}$$

$$L = \frac{\rho}{1 - \rho} \qquad \text{Exponential}$$

$$L = \frac{2k\rho - \rho^2(k - 1)}{2k(1 - \rho)} \qquad \text{Erlang}$$

$$L = \frac{\rho^2 + \rho(1 - \rho)4s(1 - s)}{4s(1 - s)(1 - \rho)} \qquad \text{Hyper-exponential}$$

In all cases, the total number of entities waiting for service and the mean times spent waiting, both for service to begin and for service to be completed, can be derived from L with the following formulae:

$$L_w = L - \rho$$

$$W = \frac{L}{\lambda} = LT_a$$

$$W_w = \frac{L_w}{\lambda} = L_w T_a$$

Figure 7-7 plots L_w, the mean number of entities waiting, against the server utilization for several cases. There are two cases for the hyper-exponential, with $k = 2$ and 4, corresponding to coefficients of variation of 1.414 and 2, respectively. There are also two cases for the Erlang distribution, also with $k = 2$ and 4, which correspond to coefficients of variation of 0.71 and 0.5, respectively. The exponential and constant service time cases are also shown, for which the coefficient of variations are 1 and 0, respectively. It can be seen that the number of entities waiting rises with increasing server utilization, the rate of increase becoming greater as a utilization of 1 is approached. The rise also starts earlier as the coefficient of

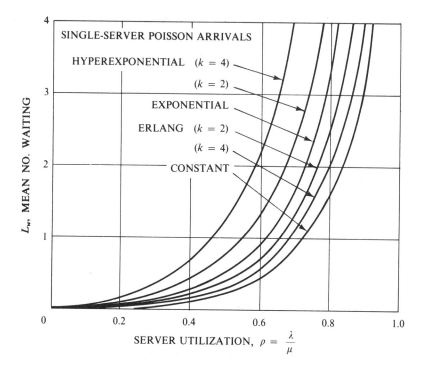

Figure 7-7. Number of entities waiting with various service time distributions.

variation gets bigger. When the utilization exceeds 1, the system will have a constantly increasing waiting line, whatever the service distribution.

Figure 7-8 shows the mean number waiting in a multiserver system with Poisson arrivals and exponential service time distribution. The case of a single server is included for comparison. Other curves are for the cases of 2, 5, and 10 servers. It can be seen that increasing the number of servers with the same server utilization (that is, the same traffic intensity per server) decreases the number waiting for service to begin.

7-13

Utilization as a Design Factor

Figures 7-7 and 7-8 have shown the number waiting as a function of the server utilization. Because of the relationships given above, the number of entities in the system and the times spent waiting can also be expressed in terms of the server utilization. In fact, most measures of interest in server systems can be expressed this way.

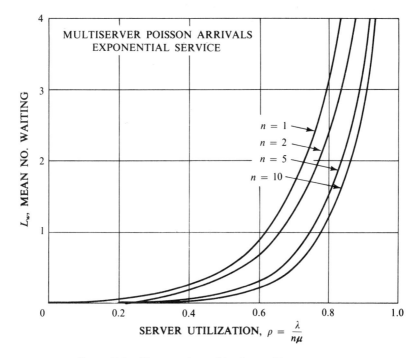

Figure 7-8. Mean number waiting in a multiserver system.

Knowing how the measures depend upon the utilization gives an understanding of how system parameters should be changed to modify performance. If the number of waiting entities is judged to be too high, for example, the utilization needs to be lowered. The definitions show this can be done in three ways: the arrival rate can be lowered, the service rate can be increased, or the number of servers can be increased. Any of these three changes, or some combination of them, will lower ρ. On the other hand, if the service is more than satisfactory, better utilization can be made of the system resources by changing the factors in the other direction.

7-14

Grade of Service

A criterion by which many systems are judged is the *grade of service*, which places some limit, or limits, on the length of time an entity must wait for service—measured either as the time for service to begin or to be completed. This is an important characteristic of any system involving human beings, for example, a telephone system, or any computer-based terminal system. The limit might be absolute, but it is usually expressed in probability terms. The requirement, for example, might be that not more than 10% of the requests for service should have to wait more than three seconds.

To meet such criteria, it is necessary to know the probability of waiting a given length of time, and so waiting time probabilities are important outputs of simulation studies. As an illustration, Fig. 7-9 shows the probability of waiting longer than a given time for service to begin, in the case of a single server with Poisson arrivals and an exponential service time distribution. The probability, denoted by $P_w(t)$, and the corresponding function for the probability of service being completed, $P(t)$ for this particular system are

$$P(t) = e^{-(1-\rho)\mu t}$$
$$P_w(t) = \rho P(t)$$

As can be seen from Fig. 7-9, the results can be plotted as a function of μt, with a separate curve for each value of ρ.

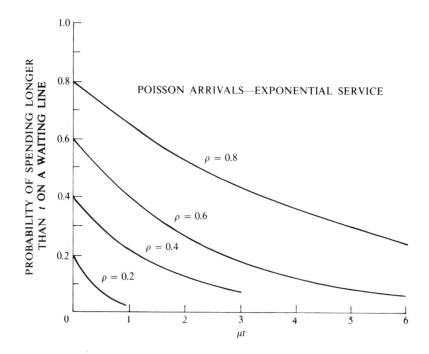

Figure 7-9. Probability of spending time waiting for service to begin.

Suppose the specified grade of service is that at least one in four requests should get immediate attention, and not more than 20% should have to wait more than 30 seconds. The first criterion means that there shall be no more than a 0.75 chance of any delay. This is decided by the points at which the curves cross the $t = 0$ axis. The condition is that p shall be no greater than 0.75, since that is the curve that crosses the axis at 0.75, and all lower utilizations give lesser values. The second criterion requires knowing the mean service time. Suppose it is 10 seconds, corresponding to $\mu = 0.1$, then a time of 30 seconds corresponds to a value of $\mu t = 3$. From Fig. 7-9, the curve for $p = 0.6$ shows a probability of about 20% for μt to be greater than 3, and, again, lesser utilizations give lower probabilities. The second criterion, therefore, requires that p be no bigger than 0.6. This is a more stringent restriction that the previous one, so it will become the design criterion set for the system. The result is that a maximum mean inter-arrival time of $10/0.6 = 16.7$ seconds must be set, so that, with a mean service time of 10 seconds, the server utilization is held to 0.6.

In the case of telephone systems, an important design criterion is that there shall be enough servers, in the form of available circuits, to avoid having any customer wait for service: it being assumed that such customers will be lost. A message switching system which does not have the ability to store delayed messages needs to meet the same stringent condition of no losses.

Short of providing an individual circuit for each customer, or message source, such perfect service cannot be guaranteed. The design can, however, be set for a given grade of service. Figure 7-10 shows the probability of loss against traffic intensity for different numbers of servers. The assumed conditions are that the arrivals are Poisson, and they come from an infinite population of sources. In this multiserver loss system, the distribution of the service time is not important: the results depend only on the mean value.

To illustrate the use of the curves, the following figures, derived by reading across the graph at the 0.01 line, show the traffic that can be carried with not more than 1% of the customers, or messages, being lost:

No. of Servers	Traffic Intensity
2	0.15
3	0.31
4	0.86
5	1.37
6	1.78
8	3.15
10	4.40

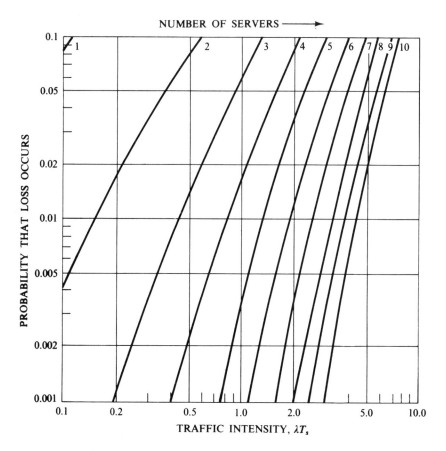

Figure 7-10.   Probability of loss in multiserver system.

The rapid rise of the traffic intensity with number of servers, demonstrates a characteristic of server systems which is of great importance to telephone systems. Doubling the number of servers from 2 to 4 raises the traffic carrying capacity, for this grade of service, by a factor of 5.7. Doubling again to 8 servers, produces another factor of 3.7. The net effect of quadrupling from 2 to 8 produces a 21-fold increase in traffic carrying capacity.

Exercises

7-1   Use Table 7-4 and the first column of Table 6-3 to compute the arrival times of the first 10 arrivals of a Poisson arrival pattern with a mean inter-arrival time of 100. Plot the results as an arrival distribution.

7-2 Calculate the probability of there being n arrivals ($n = 0, 1, \ldots, 10$) in an interval of 10 seconds when the arrivals have a Poisson distribution with a mean value of 0.4.

7-3 Use Table 7-4 and the first four columns of Table 6-3 to generate 20 numbers from a 4th order Erlang distribution with mean value 12. Plot the numbers as an arrival distribution.

7-4 Using the method described in Sec. 7-7, generate 20 numbers with a hyper-exponential distribution having $k = 4$ and a mean value of 10. Use Table 7-4 and column 1 of Table 6-3 to compute the exponentially distributed numbers, and column 2 of Table 6-3 to determine the choice of stages. Plot the results as an arrival distribution.

7-5 Use Table 7-6 and the first column of Table 6-3 to generate 20 numbers having a normal distribution with mean value 50 and standard deviation 5. Plot the distribution of the numbers.

7-6 Determine the coefficient of variation for the following data and judge what distribution best fits the data. Plot the data as an arrival distribution:

95	52	79	31	79	140	175	101	140	158
80	92	90	67	88	97	87	98	105	80
103	106	114	115	100	132	112	162	104	124
116	81	197	138	100	44	156	92	72	97
88	69	117	64	51	138	60	67	87	78

Bibliography

1 AHRENS, J. H., AND U. DIETER, "Computer Methods for Sampling from the Exponential and Normal Distributions," *Commun. ACM*, XV, no. 10 (1972), 873–882.

2 BOX, G. E. P., AND MERVIN E. MULLER, "A Note on the Generation of Random Normal Deviates," *Ann. Math. Stat.*, XXIX, (1958), 610–611.

3 BURR, IRVING W., "A Useful Approximation to the Normal Distribution With Application to Simulation," *Technometrics*, IX, no. 4 (1967), 647–651.

4 FELLER, WILLIAM, *An Introduction to Probability Theory and Its Applications*, vols. 1 and 2, New York: John Wiley & Sons, Inc., 1957 (vol. 1) and 1966 (vol. 2).

5 HAIGHT, FRANK A., *Handbook of the Poisson Distribution*, New York: John Wiley & Sons, Inc., 1967.

6 HAMMING, RICHARD W., *Numerical Methods for Scientists and Engineers*, New York: McGraw-Hill Book Company, 1962.

7   IBM Corp., *GPSS II User's Manual*, (Form SH20-6346), White Plains, N.Y.

8   ——, *2314 Reference Manual* (Form A26-3599).

9   KLEINROCK, L., *Queueing Systems, vol. 1: Theory*, New York: John Wiley & Sons, 1975.

10 MacLAREN, M. D., G. MARSAGLIA, AND T. A. BRAY, "A Fast Procedure for Generating Exponential Random Variables," *Commun. ACM*, VII, no. 5 (1964), 298–300.

11 MARSAGLIA, G., M. D. MacLAREN, AND T. A. BRAY, "A Fast Procedure for Generating Normal Random Variates," *Commun. ACM*, VII, no. 1 (1964), 4–9.

12 MARTIN, JAMES, *Design of Real-Time Computer Systems*, pp. 441–446, Englewood Cliffs, N.J.: Prentice-Hall, Inc., 1967.

13 MORSE, PHILIP M., *Queues, Inventories, and Maintenance*, New York: John Wiley & Sons, Inc., 1958.

8

DISCRETE
SYSTEM SIMULATION

8-1
Discrete Events

In Sec. 3-7 we demonstrated the general nature of the numerical computations involved in discrete system simulation, using a simple model describing the processing of documents. We now discuss discrete system simulation more fully, in terms directed toward executing the simulation with digital computers.

We saw that the model used in discrete system simulation has a set of numbers to represent the state of the system, such as the numbers of Table 3-1, representing the processing of documents. A number used to represent some aspect of the system state is called a *state descriptor*. Some state descriptors range over values that have physical significance, such as the number representing the count of documents. Others represent conditions, such as the flag denoting whether a break in work is due.

As the simulation proceeds, the state descriptors change value. We define a *discrete event* as a set of circumstances, that causes an instantaneous change in one or more system state descriptors. Implicit in the definition is the assumption that the system change is one that has been considered of sufficient importance to be represented in the model. In addition, it is possible that two different events occur simultaneously, or are modeled as being simultaneous, so that not all changes of state descriptors occurring simultaneously necessarily belong to a single event.

The term simultaneous, of course, refers to the occurence of the changes in the system, and not to when the changes are made in the model—in which, of necessity, the changes must be made sequentially. A system simulation must contain a number representing time. The simulation proceeds by executing all the changes to the system descriptors associated with each event, as the events occur, in chronological order. The way in which events are selected for execution, particularly when there are simultaneous events, is an important aspect of programming simulations. It will be discussed in Chap. 13, when discussing programming techniques used in simulation programming languages.

8-2

Representation of Time

The passage of time is recorded by a number referred to as *clock time*. It is usually set to zero at the beginning of a simulation and subsequently indicates how many units of simulated time have passed since the beginning of the simulation. Unless specifically stated otherwise, the term *simulation time* means the indicated clock time and not the time that a computer has taken to carry out the simulation. As a rule, there is no direct connection between simulated time and the time taken to carry out the computations. The controlling factor in determining the computation time is the number of events that occur. Depending upon the nature of the system being simulated, and the detail to which it is modeled, the ratio of the simulated time to the real time taken can vary enormously. If a simulation were studying the detailed workings of a digital computer system, where real events are occurring in time intervals measured in fractions of microseconds, the simulation, even when carried out by a high speed digital computer, could easily take several thousand times as long as the actual system operation. On the other hand, for the simulation of an economic system, where events have been aggregated to occur once a year, a hundred years of operation could easily be performed in a few minutes of calculations.

Two basic methods exist for updating clock time. One method is to advance the clock to the time at which the next event is due to occur. The other method is to advance the clock by small (usually uniform) intervals of time and determine at each interval whether an event is due to occur at that time. The first method is referred to as *event-oriented*, and the second method is said to be *interval-oriented*. Discrete system simulation is usually carried out by using the event-oriented method, while continuous system simulation normally uses the interval-oriented method.

It should be pointed out, however, that no firm rule can be made about the way time is represented in simulations for discrete and continuous systems. An interval-oriented program will detect discrete changes and can therefore simulate discrete systems, and an event-oriented program can be made to follow continuous changes by artificially introducing events that occur at regular time intervals. Further, the event-oriented method is not necessarily faster than the interval-oriented method for discrete systems, (7).

Another approach to representing the passage of time has been called *significant event simulation*, (1). The method is applicable to continuous systems in which there are quiescent periods. The interval between events in the event-oriented approach is, of course, a quiescent period, but it involves having the model's representation of the system activities create a notice of the event that terminates the interval. The significant event approach assumes that simple analytic functions, such as polynomials of low order, can be used to project the span of a quiescent period. The event that ends the period could be any one of several alternatives, each of which has a projected span. The significant event is the one with the least span. Determining this event, by simple comparisons of the projections, allows the clock to be updated by an extended period of time; achieving the same thing would otherwise have cost the effort of executing the updating of many fruitless intervals of fixed size.

An example, which is essentially the one used in Ref. (1), is an automobile traveling at constant acceleration. Its movement might result in a significant event for several reasons: the automobile might approach the preceding vehicle closer than a specific limit, it might reach the end of the road, its velocity might reach some limit, or it might come to rest. If the initial conditions are known, the elapsed time for each of these possible events can be calculated from simple formulae. Conditions relating to all other vehicles must, of course, also be evaluated before deciding on the next significant event for the entire system.

The method includes, however, selecting the next significant event, and not just its time of occurrence. As such, it is more than a timing mechanism: it should more properly be described as an event-scanning method. It is, in fact, closely related to the activity-oriented scanning method, discussed in Chap. 13, which describes event-scanning methods.

8-3

Generation of Arrival Patterns

An important aspect of discrete system simulation is the generation of exogenous arrivals. It is possible that an exact sequence of arrivals has been specified for the simulation. For example, discrete system simulation is extensively used for testing the design of logic circuits, such as components of digital computers. A particular sequence of signals might be designed as the simulation input to see if the design reacts as expected.[1]

The sequence of inputs may have also been generated from observations on a system. In particular, computer system designs, and especially the programming components of the system, are often tested with a record, gathered from a running system, that is representative of the sequence of operations the computer system will have to execute. This approach has been called *trace driven simulation*, (3) and (9). Program monitors can be incorporated with, or attached to, the running system to extract the data with little or no disturbance of the system operation (10).

When there is no interaction between the exogenous arrivals and the endogenous events of the system, it is permissible to create a sequence of arrivals in preparation for the simulation. Usually, however, the simulation proceeds by creating new arrivals as they are needed.

The exogenous arrival of an entity is defined as an event and the arrival time of the next entity is recorded as one of the event times. When the clock time reaches this event time, the event of entering the entity into the system is executed, and the arrival time of the following entity is immediately calculated from the inter-arrival time distribution. The term *bootstrapping* is often used to describe this process of making one entity create its successor. The method requires keeping only the arrival time of the next entity; it is, therefore, the preferred method of generating arrivals for computer simulation programs.

The arriving entity usually needs to have some attribute values generated, in which case, attention must be paid to the time at which the values are generated. They could be generated at the time the arrival time is calculated, or they could be generated when the entity actually arrives. If there is no interaction between the attributes and the events occurring within the system, the generation may be done at either time. If, however, the attribute values depend upon the state of the system, it must be remembered that, at the time of generating the arrival time, the actual arrival is still an event in the future. The generation of the attribute values must then be deferred until the arrival event is executed. For example, a simulation is discussed in the next section in which telephone calls are generated. The call length and the origin of the call need to be generated. There is no interaction between the distribution of call length and the state of the system, so the call length can be generated at the time the arrival time is decided or when the call arrives. However, a call cannot come from a line that is already busy, so the choice of origin must be left until the call arrives. To select the origin when the arrival time is decided

---

[1]This assumes the logic circuits are being treated as precise logical devices. There are also simulation programs that treat the circuits as continuous devices, in order to see how the signal waveforms deviate from the ideal shapes needed for precise logical devices. See, for example, (5).

carries the risk that some other call will have made the proposed origin busy before the call arrives.

8-4

Simulation of a Telephone System

To illustrate the principles involved in the simulation of a discrete system, consider the example of a simple telephone system illustrated in Fig. 8-1, (2). The system has a number of telephones (only the first eight are shown), connected to a

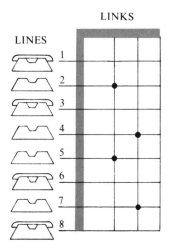

Figure 8-1. Simple telephone system.

switchboard by lines. The switchboard has a number of links which can be used to connect any two lines, subject to the condition that only one connection at a time can be made to each line. It will be assumed that the system is a lost-call system, that is, any call that cannot be connected at the time it arrives is immediately abandoned. A call may be lost because the called party is engaged, in which case the call is said to be a busy call; or it may be lost because no link is available, in which case it is said to be a blocked call. The object of the simulation will be to process a given number of calls and determine what proportion are successfully completed, blocked, or found to be busy calls.

The current state of the system, shown in Fig. 8-1, is that line 2 is connected to line 5, and line 4 is connected to line 7. One way of representing the state of the system is shown in Fig. 8-2. Each line is treated as an entity, having its availability as an attribute. A table of numbers is established to show the current status of each

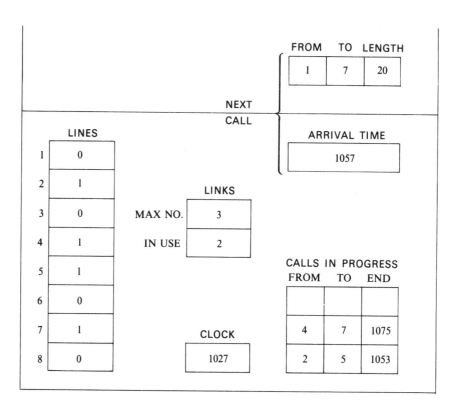

Figure 8-2. System state-1.

line. A *zero* in the table means the line is free, while a *one* means that it is busy. It is not necessary that a detailed history be kept of each individual link, since each is able to service any line. It is only necessary to incorporate in the model the constraint imposed by the fact that there is a fixed number of links (in this case, three). Under these circumstances, the group of links is represented as a single entity, having as attributes the maximum number of links and the number currently in use. Two numbers, therefore, represent the links.

To keep track of events, a number representing clock time is included. Currently, the clock time is shown to be 1027, where the unit of time is taken to be 1 second. The clock will be updated to the next most imminent event as the simulation proceeds. Each call is a separate entity having as attributes its origin, destination, length, and the time at which the call finishes. There is a list of *calls-in-progress* showing which lines each call connects and the time the call finishes. To generate the arrival of calls, the bootstrap method described in the previous section is used, so that a record is kept of the time the next call is due to arrive. It will be assumed that the call is equally likely to come from any line that is not busy at the time of arrival, and that it can be directed to any line, other than itself, irrespective of whether that line is busy or not. The choice of origin must be left until the call arrives. For convenience, the origin, destination, and call length will all be generated at that time. To help explain the action of the simulation, these choices are shown in Fig. 8-2, although, in practice, they would not have been generated until the clock reaches 1057, the time of the next arrival. The generation of the call length can, in fact, be deferred until not only has the call arrived but it is determined that it can be connected.

The set of numbers within the main box of Fig. 8-2 records the state of the system at time 1027. There are two activities causing events: new calls can arrive and existing calls can finish. As shown in Fig. 8-2, there are three future events: the call between lines 2 and 5 is due to finish at time 1053, the call between lines 4 and 7 is due to finish at time 1075, and a new call is due to arrive at time 1057. The next call, when it arrives, will come from line 1 and attempt to go to line 7. It is due to last 20 seconds.

The simulation proceeds by executing a cycle of steps to simulate each event. The first step is to *scan* the events to determine which is the next potential event. In this case, the next potential event is at 1053. The clock is updated, and the second step is to *select* the activity that is to cause the event. In this case, the activity is to disconnect a call. There are no conditions to be met when a call is disconnected, so the event will be executed; but, in general, the third step is to *test* whether the potential event can be executed. The fourth step is to *change* the records to reflect the effects of the event. The call is shown to be disconnected by setting to zero the numbers in the lines table for lines 2 and 5, reducing the number of links in use by 1, and removing the finished call from the calls-in-progress table. As a fifth and final step in the execution cycle, it may be necessary to *gather* some statistics for the simulation output. Counters are set aside to record the number of calls that have been processed, completed, or lost through being blocked or busy. With the disconnection of a call, the counts of processed calls and of completed calls are both increased by one. The state of the system then appears as shown in Fig. 8-3. Assuming the simulation is to continue, the cycle of actions just described is repeated.

It can be seen that the next event is the arrival of a call at time 1057. The clock is updated to the arrival time and the attributes of the new arrival are generated. Since the selected activity is to connect a call, it is necessary to carry

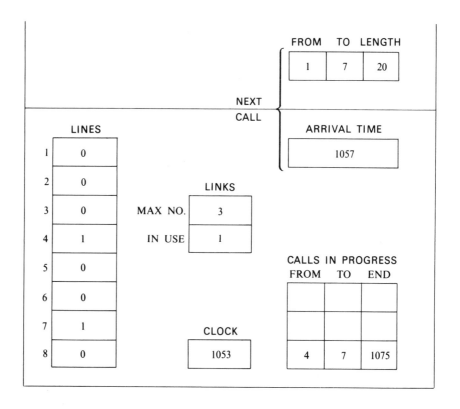

Figure 8-3. System state-2.

out tests: first to see if a link is available, then to see if the party is busy. In this case, the called party, line 7, is busy so the call is lost. Both the processed calls and the busy calls counters are increased by one. A new arrival is generated and the state of the system at the time the call is lost then appears as shown in Fig. 8-4. Suppose the next arrival time is 1063, and that when this arrival occurs the call will be from line 3 to 6 and will last 98 seconds. Again, the next potential event is an arrival, but this time the arriving call can be connected, so the state of the system moves to that shown in Fig. 8-5.

The procedure will be repeated until some limit on the length of the simulation is reached. Typically, the simulation will be run until a given number of calls has been processed or until a certain time has elapsed.

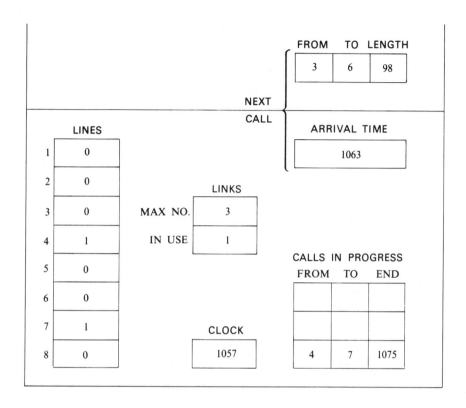

Figure 8-4. System state-3.

8-5

Delayed Calls

Suppose the telephone system is modified so that calls that cannot be connected are not lost. Instead, they wait until they can be connected. This is not characteristic of a normal telephone system involving human beings talking to each other. However, it can happen to messages in a switching system that has store-and-forward capability. We will continue, however, to describe the transactions as calls.

To keep records of the delayed calls, it is necessary to build another list like the calls-in-progress list. Recomputing the simulation, the system moves through

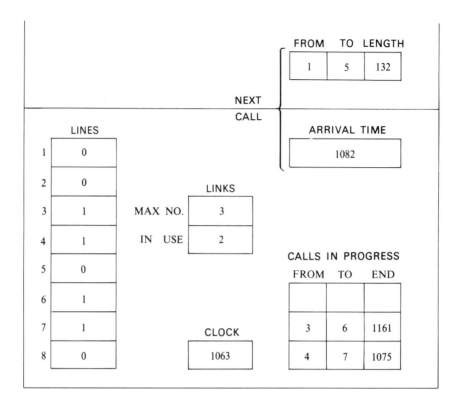

Figure 8-5. System state-4.

the first two states exactly as before. The state of the system at time 1057, when the call from line 1 has arrived, now appears as shown in Fig. 8-6, which is labelled as system state-3A. It is the same as state-3, shown in Fig. 8-4, except that it now shows the delayed call. (Although calls are not now lost, the counters will still be kept to show how many calls were delayed by encountering busy or blocked conditions.)

Now, when a call is completed, it is necessary to check the delayed call list to see if a waiting call can be connected. The next event, however, is the arrival of a call, from line 3 at time 1063, which is going to line 6. The system goes to state-4A, shown in Fig. 8-7, which is the same as the previous state-4 except that it has the record of the delayed call.

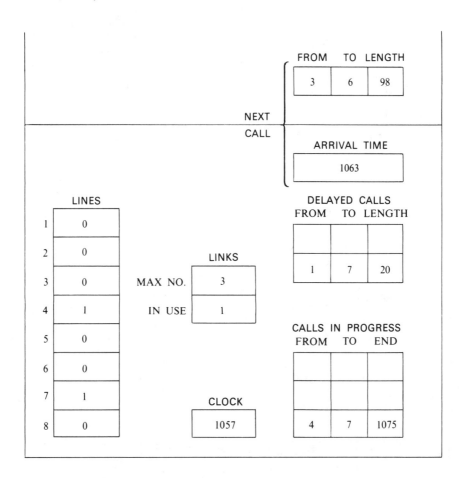

Figure 8-6. System state-3A.

We now move the system to state-5A, which reflects the changes that occur at time 1075 when the call from line 4 to line 7 ends. The records must be changed to show the removal of that call and the connection of the call that was waiting, including the transfer of the call from the delayed list to the calls-in-progress list. The result, when all changes have been made, is as shown in Fig. 8-8. Notice that the call that just arrived finishes before the call between lines 3 and 6 which arrived earlier but is due to last much longer than the new call. Since the calls-in-progress

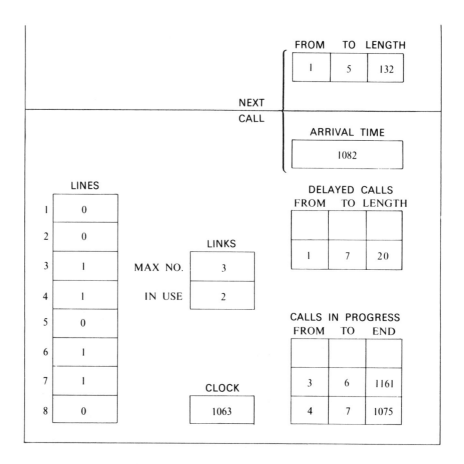

Figure 8-7. System state-4A.

table is being kept in chronological order by completion time, the new call goes ahead of the earlier call in the list.

If a computer program carries out the changes, the change to state-5A might be regarded as one complex event; or it might be regarded as two coincident events: one to disconnect a call, and a second to connect a call. The choice depends upon the program's design, a topic to be discussed in Chap. 13.

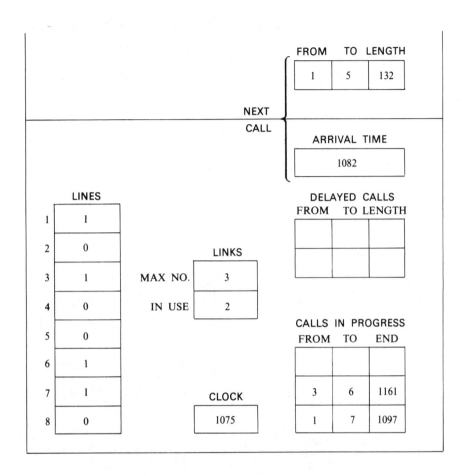

Figure 8-8. System state-5A.

8-6

Simulation Programming Tasks

Having demonstrated with a particular example the way a discrete simulation proceeds, we can now outline in general the tasks involved in preparing a computer program for a simulation. There are three main tasks to be performed, as shown in Fig. 8-9. The first is to generate a model and initialize it. From the description of the system, a set of numbers must be created to represent the state of the system. This set of numbers will be called the *system image* since its purpose is to reflect the state of the system at all times. The activities of the system must be represented as routines that create the discrete events making the changes to the system image. The second task is to program the procedure that executes the cycle of actions involved in carrying out the simulation. This procedure is referred to as the *simulation algorithm*. The third task is the generation of an output report. The statistics gathered during the simulation will usually be organized by a report generator.

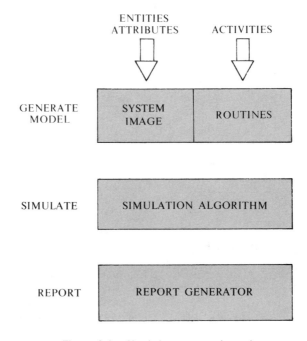

Figure 8-9. Simulation programming tasks.

The general flow of control during the execution of a simulation program is illustrated in Fig. 8-10. Shown at the top of the figure is the task of generating the model, which is executed once. At the bottom is the report generation task, which is usually executed once at the end of the simulation, although it is not unusual to print out intermediate results as the simulation proceeds. Carrying out the simulation algorithm involves repeated execution of the five steps described previously. These steps are:

1. *Find* the next potential event.
2. *Select* an activity.

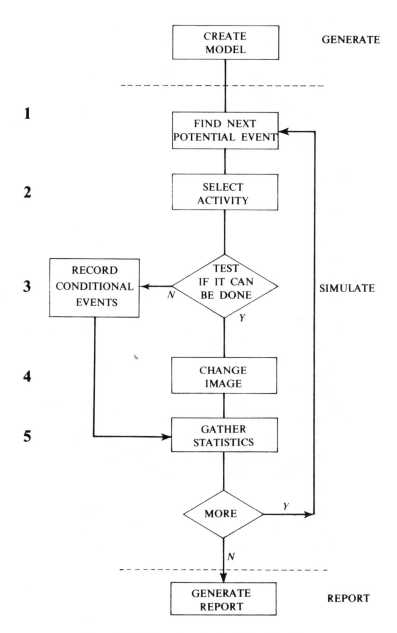

Figure 8-10. Execution of a simulation algorithm.

3. *Test* if the event can be executed.
4. *Change* the system image.
5. *Gather* statistics.

If the test made at step three finds that the event due to occur cannot, in fact, be executed, the event may be abandoned, as happened in the lost-call version of the telephone system. If, however, the event is to be executed later, when conditions are suitable, it is said to become a *conditional event*. As indicated in Fig. 8-10, some record must be made of the conditional events. The existence of the conditional events must be considered when finding the next event.

8-7

Gathering Statistics

Most simulation programming systems include a report generator to print out statistics gathered during the run. The exact statistics required from a model depend upon the study being performed, but there are certain commonly required statistics which are usually included in the output. Among the commonly needed statistics are:

(a) *Counts* giving the number of entities of a particular type or the number of times some event occurred.

(b) *Summary measures*, such as extreme values, mean values, and standard deviations.

(c) *Utilization*, defined as the fraction (or percentage) of time some entity is engaged.

(d) *Occupancy*, defined as the fraction (or percentage) of a group of entities in use on the average.

(e) *Distributions* of important variables, such as queue lengths or waiting times.

(f) *Transit times*, defined as the time taken for an entity to move from one part of the system to some other part.

When there are stochastic effects operating in the system, all these system measures will fluctuate as a simulation proceeds, and the particular values reached at the end of the simulation are taken as estimates of the true values they are designed to measure. Deciding upon the accuracy of the estimates is a problem that will be taken up in Chap. 14. For the time being, we discuss the methods used to derive the estimates.

8-8

Counters and Summary Statistics

Counters are the basis for most statistics. Some are used to accumulate totals; others record current values of some level in the system. The telephone system simulation, for example, used counters to record the total number of lost and busy calls, as well as to keep track of how many links were in use at any time. Maxima or minima are easily obtained. Whenever a new value of a count is established, it is compared with the record of the current maximum or minimum, and the record is changed when necessary.

The mean of a set of N observations $x_r (r = 1, 2, \ldots, N)$ is defined as

$$m = \frac{1}{N} \sum_{r=1}^{n} x_r$$

A mean, therefore, is derived by accumulating the total value of the observations, and also accumulating a count of the number of observations. The division of the two numbers is performed at the end of the simulation run in preparation for the final output of the program.

A standard deviation was defined previously, [in Eq. (6-5)] as

$$s = \left\{ \frac{1}{(N-1)} \sum_{r=1}^{N} (m - x_r)^2 \right\}^{1/2}$$

The common method of summing the squares is based on the following expansion:

$$\sum_{r=1}^{N} (m - x_r)^2 = \sum_{r=1}^{N} x_r^2 - Nm^2$$

The accumulated sum of the observations must be kept to derive the mean value, and so the only additional record needed to derive a standard deviation is the sum of the squares. It can take large values, especially when many observations are made. Then, double precision calculations may be needed, particularly if the calculations are carried out in integer arithmetic. For a discussion of methods for calculating means and standard deviations when extremely high precision is needed, see Ref. (8).

8-9

Measuring Utilization and Occupancy

A common requirement of a simulation is measuring the load on some entity, such as an item of equipment. The simplest measure is to determine what fraction of the time the item was engaged during the simulation run. The term *utilization* will be used to describe this statistic. Typically, the time history of the equipment usage might appear as shown in Fig. 8-11. To measure the utilization, it is necessary to keep a record of the time t_b at which the item last became busy. When the item

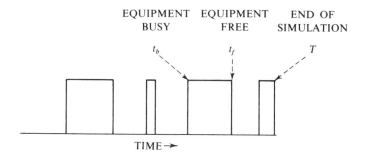

Figure 8-11. Utilization of equipment.

becomes free at time t_f, the interval $t_f - t_b$ is derived and added to a counter. At the end of the simulation run, the utilization U is derived by dividing the accumulated total by the total time T, so that, if the entity is used N times:

$$U = \frac{1}{T} \sum_{r=1}^{N} (t_f - t_b)_r$$

A discrete simulation program, updating time as events occur, will measure the intervals $t_f - t_b$ directly. A continuous simulation program updating time in small intervals will need to build up the count by counting the number of intervals in which the item is busy.

Note that it is important to check whether the item is busy at the end of the run, and, if so, add to the counter a quantity representing the engagement from the last time it became busy to the end of the simulation. Correspondingly, it is also important to check the initial conditions to see if the entity is busy at the beginning of the run.

In dealing with groups of entities, rather than individual items, the calculation is similar, requiring that information about the number of entities involved also be kept. Figure 8-12, for example, represents, as a function of time, the number of links in a telephone system that are busy. To find the average number of links in use, a record must be kept of the number of links currently in use and the time at which the last change occurred. If the number changes at time t_r to the value n_r, then, at the time of the next change t_{r+1}, the quantity $n_r(t_{r+1} - t_r)$ must be calculated and added to an accumulated total. The average number in use during the simulation run, A, is then calculated at the end of the run by dividing the total by the total simulation time T, so that:

$$A = \frac{1}{T} \sum_{r=1}^{N} n_r(t_{r+1} - t_r)$$

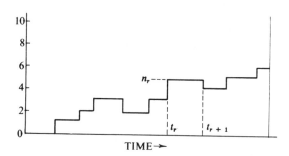

Figure 8-12. Time history of busy telephone links.

Figure 8-12 might also represent the number of entities waiting on a queue, in which case, the calculation gives the mean number of entities waiting.

If there is an upper limit on the number of entities, as there was a limit on the number of links in the telephone system, the term *occupancy* is often used to describe the average number in use as a ratio to the maximum. Thus, if there are M links in a telephone exchange and the quantity n_r is the number busy in the interval t_r to t_{r+1}, the average occupancy, assuming the number n_r changes N times, is

$$B = \frac{1}{NM} \sum_{r=1}^{N} n_r(t_{r+1} - t_r)$$

As mentioned before, it is important to check the conditions at the beginning and end of the simulation.

An important difference between utilization and occupancy statistics, as they have been defined here, is that for utilization, timing information must be kept for each individual entity. Occupancy statistics only require keeping a count of a class of entities, and recording the last time the count changed, just two numbers. If the number of active entities is large, it can cost a great deal more space and time to record utilizations than occupancies.

8-10

Recording Distributions and Transit Times

Determining the distribution of a variable requires counting how many times the value of the variable falls within specific intervals. A table sets aside locations in which to record the values defining the intervals and to accumulate each count. As each new observation is made, its value is compared with the limits established for the intervals, and 1 is added to the counter for one interval.

Normally, the tabulation intervals are uniform in size and the specifications will be for

(a) The lower limit of tabulation,
(b) The interval size,
(c) The number of intervals.

The meaning of these terms is illustrated in Fig. 8-13. The user will not always gauge accurately what the potential range of values will be, so it is customary to count how many times an observation falls below the lower limit and beyond the upper limit.

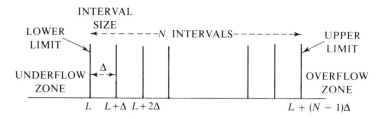

Figure 8-13.   Definition of a distribution table.

It is customary to accumulate the number of observations and the sum of the squares, in order to calculate the mean value and the standard deviation, at the same time the distribution is being derived. Each observation, x_i, will therefore result in a count of 1 being added to the appropriate counter, the addition of x_i to the accumulated total $\sum x_i$, and the addition of x_i^2 to the sum $\sum x_i^2$.

Figure 8-14 illustrates the space that needs to be set aside. Note that, while the distribution derived is an approximation, since it matches the values of the observations to an interval, the mean or standard deviation will be accurate within the accuracy limits of the computer, even if some observations fall outside the table limits.

The nature of the random variable being measured determines when the observations are made. If the mean waiting time for a service were to be measured, an observation would be taken as each entity starts to receive service, so that the times at which the observations are tabulated, or recorded for tabulation, are randomly spaced. To measure the distribution of the number of entities waiting, however, observations would be taken at uniform intervals of time.

In addition to deriving the mean and standard deviation of a distribution, the final output will often express the data in some other convenient form. For example, the cumulative distribution may be given, or the distribution may be rescaled to express the counts as percentages of total observations or to express the interval size in terms of the mean value of the observations. All these derived results can

	L, LOWER LIMIT	UNDERFLOW COUNT
$i = 1$		
	$L + \Delta$	1st INTERVAL COUNT
	$L + i\Delta$	ith INTERVAL COUNT
$i = N - 1$	$L + (N - 1)\Delta$	$(N - 1)$th INTERVAL COUNT
$i = N$		OVERFLOW COUNT

NUMBER OF ENTRIES
TOTAL OF ENTRIES
SUM OF SQUARES

Figure 8-14.   Space reserved for a distribution table.

be calculated at the end of the run. A typical example of such a table printout will be found in Fig. 9-9.

To measure transit times, the clock is used in the manner of a time stamp. When an entity reaches a point from which a measurement of transit time is to start, a note of the time of arrival is made. Later, when the entity reaches the point at which the measurement ends, a note of the clock time upon arrival is made and compared with the first time to derive the elapsed interval.

8-11

Discrete Simulation Languages

A number of programming languages have been produced to simplify the task of writing discrete system simulation programs. A list of 23 such languages will be found in (4).

Essentially, these programs embody a language with which to describe the system, and a programming system that will establish a system image and execute a simulation algorithm. Each language is based upon a set of concepts used for describing the system. The term *world-view* has come to be used to describe this aspect of simulation programs (6). The user of the program must learn the world-view of the particular language he is using and be able to describe the system in those terms. Given such a description, the simulation programming system is able to establish a data structure that forms the system image. It will also compile, and sometimes supply, routines to represent the activities. Routines are supplied to carry out such functions as scanning events, updating the clock, gathering statistics, and maintaining events in time and priority sequence. These are needed to effect the simulation algorithm. Most programs also provide a report generator.

There is, however, a great variety in both the world-views of the languages and the degree to which the programming systems relieve the user of programming details. In general, most languages view the world in terms of entities with attributes and activities, as these concepts were introduced in Chap. 1.

It is not feasible to discuss here all the simulation languages that are available. Instead, the discussion will be limited to two languages, GPSS and SIMSCRIPT. The next two chapters are devoted to GPSS, and the following two chapters are given to SIMSCRIPT. The reasons for choosing these two specific languages are that they are among the most widely used languages, and they illustrate the divergence in design considerations. As will be discussed in Chap. 13, they also represent two major approaches to the problem of implementing a simulation algorithm.

GPSS has been written specifically for users with little or no programming experience, while SIMSCRIPT, along with many other simulation languages, requires programming skill to the level where the user is able to program in FORTRAN or ALGOL. The simplifications of GPSS result in some loss of flexibility; so that, while SIMSCRIPT requires more programming skill, it is capable of representing more complex data structures and can execute more complex decision rules. Both GPSS and SIMSCRIPT appear to be general enough to be equally applicable to a wide variety of systems. The differences are summed up by saying that the greater programming flexibility of SIMSCRIPT means that, in more complex models, SIMSCRIPT is able to produce a more compact model that requires less storage space and, generally, will be executed more rapidly.

Exercises

8-1 Recalculate the telephone system simulation, assuming the inter-arrival time is exponentially distributed and the call lengths are normally distributed. The mean inter-arrival time is 20 and the calls have a mean length of 80 with a standard deviation of 20. Use Table 7-4 and column 1 of Table 6-3 to generate the call arrivals, and use Table 7-6 and column 2 of Table 6-3 to generate the call lengths. Assume the system starts with a call from line 2 to line 5, due to end at time 70, and a call from 4 to 7 to end at 150. Assume that there are 8 lines and 3 links. The origins and destinations of the first five calls are:

Call Number	Origin	Destination
1	1	6
2	3	8
3	1	8
4	5	1
5	2	5

How many of the first five calls are blocked or find busy lines?

8-2 Consider the market model of Sec. 3-9 for the stable conditions (case a). Suppose the market is not cleared; instead, the supplier only sells $N\%$ of his goods, where N is a uniformly distributed random number between 80 and 99, inclusive. The price he receives is determined by the point of the demand curve that corresponds to the amount sold. The supplier continues to determine the supply from the last market price. Use the random numbers in column 1 of Table 6-3 to simulate the first 6 periods of the market.

8-3 The birth rate of a population is normally distributed with mean value 5% and standard deviation 0.5%. The death rate is also normally distributed with mean 6% and standard deviation 1%. Use Table 7-6 and columns 1 and 2 of Table 6-3 to compute births and deaths, respectively, and calculate the population changes for 5 years at 1 year intervals.

8-4 A reservoir holds 1,000,000 gallons of water, and it is drained at a steady rate of 10,000 gallons a day. Rainfall occurs with a Poisson distribution with a mean rate of 1 in 10 days. The amount of rain captured by the reservoir is normally distributed with a mean of 80,000 gallons and a standard deviation of 5,000 gallons. Use Tables 7-4 and 7-6 and columns 3 and 4, respectively, of Table 6-3, to compute the time and quantity of rainfalls. Find the contents of the reservoir after 60 days. Assume the reservoir starts full and that water exceeding the capacity of the reservoir is lost.

8-5 Cars arrive randomly at a toll booth and pay tolls. If necessary, they wait in a queue to be served in order of arrival. The inter-arrival time, measured to the nearest second, is found to be uniformly distributed between 0 and 9, inclusive. The time to pay is also random and between 0 and 9 seconds, but with the following distribution:

$$f(t) = \frac{t^{-1/2}}{6} \qquad 0 \le t \le 9$$

$$= 0 \qquad \text{elsewhere}$$

Generate the arrival times from column 5 and the paying times from column 6 of Table 6-3. Truncate all derived numbers to the nearest integer. Find the arrival times and the times taken to pay the tolls for the first five cars. At what time does the fifth car clear the toll booth?

Bibliography

1 Babich, Alan F., John Grason, and David L. Parnas, "Significant Event Simulation," *Commun. ACM*, XVIII, no. 6 (1975), 323–329.

2 Blake, K., and G. Gordon, "System Simulation With Digital Computers," *IBM Syst. J.*, III, no. 1 (1964), 14–20.

3 Cheng, P. S., "Trace-Driven System Modeling," *IBM Syst. J.*, VIII, no. 4 (1969), 280–289.

4 Fishman, George S., *Concepts and Methods in Discrete Event Digital Simulation*, pp. 276–278, New York: John Wiley & Sons, Inc., 1973.

5 Jensen, R. W., and M. D. Lieberman, *IBM Electronic Circuit Analysis Program: Techniques and Applications*, Englewood Cliffs, N. J.: Prentice-Hall, Inc., 1968.

6 Krasnow, Howard S., and Reino Merikallio, "The Past, Present, and Future of General Simulation Languages," *Manage. Sci.*, XI, no. 2 (1964), 236–267.

7 Nance, Richard E., "On Time Flow Mechanisms for Discrete System Simulation," *Manage. Sci.: Theory*, XVIII, no. 1 (1971), 59–73.

8 Neeley, Peter M., "Comparison of Several Algorithms for Computation of Means, Standard Deviations and Correlation Coefficients," *Commun. ACM*, IX, no. 7 (1966), 496–499.

9 Sherman, Stephen W., "Trace Driven Modeling: An Update," *Proc. Symposium on Simulation of Computer Systems IV*, National Bureau of Standards, Boulder, Colorado, 1976, pp. 87–91. (Available from ACM, 1133 Ave. of the Americas, N. Y., N. Y., 10036.)

10 ———, Forest Baskett III, and J. C. Browne, "Trace-Driven Modeling and Analysis of CPU Scheduling in a Multiprogramming System," *Commun. ACM*, XV, no. 12 (1972), 1063–1069.

9

INTRODUCTION
TO GPSS

9-1

GPSS Programs

The General Purpose Simulation System language[1] has been developed over many years, principally by the IBM Corporation. Originally published in 1961 (2), it has evolved through several versions (1), (5), (4), to the latest version, (3). It has been implemented on several different manufacturers' machines, and there are variations in the different implementations. With regard to the successive IBM versions of the language, all but GPSS/360 and GPSS V, are obsolete. Of the two current versions, GPSS/360 models can operate with GPSS V, with some minor exceptions and modifications. GPSS V, however, is more powerful and has more language statements and facilities. The differences are described in Appendix 7 of (3). If these extensions are avoided, GPSS V models will run under GPSS/360.

The description given in this and the next chapter will be of GPSS V as implemented by the IBM Corporation, (6), (7), (8). Some of the more significant differences between GPSS/360 and GPSS V will be noted. For simplicity, the language will be called GPSS, except where comparisons are being made.

---

[1] In earlier versions of the program, GPSS stood for General Purpose System Simulation.

9-2

General Description

The system to be simulated in GPSS is described as a block diagram in which the blocks represent the activities, and lines joining the blocks indicate the sequence in which the activities can be executed. Where there is a choice of activities, more than one line leaves a block and the condition for the choice is stated at the block.

The use of block diagrams to describe systems is, of course, very familiar. However, the form taken by a block diagram description usually depends upon the person drawing the block diagram. To base a programming language on this descriptive method, each block must be given a precise meaning. The approach taken in GPSS is to define a set of 48 specific *block types*, each of which represents a characteristic action of systems. The program user must draw a block diagram of the system using only these block types.

Each block type is given a name that is descriptive of the block action and is represented by a particular symbol. Figure 9-1 shows the symbols used for the block types that will be described in this and the next chapter. To assist the reader, the block diagrams drawn in this chapter will name the block types, although this is not usually done by a programmer familiar with GPSS. Coding instructions for all the described block types are similarly brought together in Table 9-1. Table 9-2 describes the control statements. Each block type has a number of data fields. As the blocks are described, the fields will be referred to as field A, B, C, and so on, reflecting the order in which they are specified.

Moving through the system being simulated are entities that depend upon the nature of the system. For example, a communication system is concerned with the movement of messages, a road transportation system with motor vehicles, a data processing system with records, and so on. In the simulation, these entities are called *transactions*. The sequence of events in real time is reflected in the movement of transactions from block to block in simulated time.

Transactions are created at one or more **GENERATE** blocks and are removed from the simulation at **TERMINATE** blocks. There can be many transactions simultaneously moving through the block diagram. Each transaction is always positioned at a block and most blocks can hold many transactions simultaneously. The transfer of a transaction from one block to another occurs instantaneously at a specific time or when some change of system condition occurs.

A GPSS block diagram can consist of many blocks up to some limit prescribed by the program (usually set to 1,000). An identification number called a *location* is given to each block, and the movement of transactions is usually from one block to the block with the next highest location. The locations are assigned automatically by an assembly program within GPSS so that, when a problem is coded, the blocks

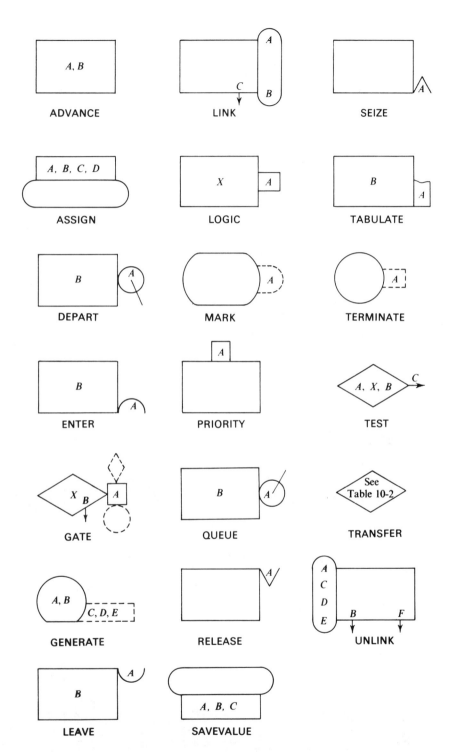

Figure 9-1. GPSS block-diagram symbols.

TABLE 9-1 GPSS Block Types

Operation	A	B	C	D	E	F
ADVANCE	Mean	Modifier				
ASSIGN	Param. No. (±)	Source	(Funct. No.)	Param. type		
DEPART	Queue No.	(Units)				
ENTER	Storage No.	(Units)				
GATE†	Item No.	(Next block B)				
GENERATE	Mean	Modifier	(Offset)	(Count)	(Priority)	(Params.)
LEAVE	Storage No.	(Units)				
LINK	Chain No.	Order	(Next block B)			
LOGIC {R S I}	Switch No.					
MARK	(Param. No.)					
PRIORITY	Priority					
QUEUE	Queue No.	(Units)				
RELEASE	Facility No.					
SAVEVALUE	S.V. No. (±)	SNA				
SEIZE	Facility No.					
TABULATE	Table No.					
TERMINATE	(Units)					
TEST†	Arg. 1	Arg. 2	(Next block B)			
TRANSFER†	Select. Factor	Next block A	Next block B			
UNLINK	Chain No.	Next block A	Count	(Param. No.)	(Arg.)	(Next block B)

†Columns 13 and 14. See Sec. 10-7 for codes.
( ) indicates optional field.

TABLE 9-2 GPSS Control Statements

Location	Operation	A	B	C	D
	CLEAR				
	END				
Function No.	FUNCTION	Argument	$\left\{\begin{matrix}C\\D\\L\end{matrix}\right\}$ No. of Points		
	INITIAL	Entity	Value		
	JOB				
	RESET				
	SIMULATE				
	START	Run Count	(NP)		
Storage No.	STORAGE	Capacity			
Table No.	TABLE	Argument	Lower limit	Interval	No. of intervals

are listed in sequential order. Blocks that need to be identified in the programming of problems (for example, as points to which a transfer is to be made) are given a symbolic name. The assembly program will associate the name with the appropriate location. Symbolic names of blocks and other entities of the program must be from three to five non-blank characters of which *the first three must be letters.*

9-3
Action Times

Clock time is represented by an integral number, with the interval of real time corresponding to a unit of time chosen by the program user. The unit of time is not specifically stated but is implied by giving all times in terms of the same unit. One block type called ADVANCE is concerned with representing the expenditure of time. The program computes an interval of time called an *action time* for each transaction as it enters an ADVANCE block, and the transaction remains at the block for this interval of simulated time before attempting to proceed. The only other block type that employs action time is the GENERATE block, which creates transactions.[2] The action time at the GENERATE block controls the *interval* between successive arrivals of transactions.

The action time may be a fixed interval (including zero) or a random variable, and it can be made to depend upon conditions in the system in various ways. An

[2]A block type called SPLIT, which will not be described in detail, is the only other block type that can create transactions. It creates copies of transactions that enter the block.

action time is defined by giving a *mean* and *modifier* as the A and B fields for the block. If the modifier is zero, the action time is a constant equal to the mean. If the modifier is a positive number ($\leq$ mean), the action time is an integer random variable chosen from the range mean $\pm$ modifier, with equal probabilities given to each number in the range. Sometimes, this uniform distribution is an accurate representation of a random process in the system, but the principal purpose in providing this way of representing a random time is to allow for cases where randomness is known to exist but no detailed information is available on the probability distribution.

It is possible to introduce a number of *functions*, which are tables of numbers relating an input variable to an output variable. Details of the functions are given in Sec. 10-3. By specifying the modifier at an ADVANCE or GENERATE block to be a function, the value of the function controls the action time. The action time is derived by *multiplying* the mean by the value of the function. Various types of input can be used for the functions, allowing the functions to introduce a variety of relationships among the variables of a system. In particular, by making the function an inverse cumulative probability distribution, and using as input a random number which is uniformly distributed, the function can provide a stochastic variable with a particular nonuniform distribution in the manner described in Sec. 6-10.

The GENERATE block normally begins creating transactions from zero time, and continues to generate them throughout the simulation. The C field, however, can be used to specify an offset time as the time when the first transaction will arrive. If the D field is used, it specifies a limit to the total number of transactions that will come from the block.

As will be explained in Sec. 10-1, transactions have a priority level and they can carry items of data called *parameters*. The E field determines the priority of the transactions at the time of creation. If it is not used, the priority is of the lowest level.

Parameters can exist in four formats. They can be signed integers of fullword, halfword, or byte size, or they can be signed floating-point numbers. If no specific assignment of parameter type is made, the program creates transactions with 12 halfword parameters. Any number of parameters (including zero) of any type, up to a limit of 255, can be specified by using fields F, G, H, and I of the GENERATE block. The symbol nPx will call for *n* parameters of type *x*, where *x* takes the value F, H, B, or L for fullword, halfword, byte size, and floating-point, respectively.[3]

For the examples used here, there are no special requirements for parameter types. The default case of 12 halfword parameters, therefore, will be used. The C, D, and E fields, also, will not be needed. In such cases, where only the A and B

[3]GPSS/360 allows up to 100 parameters of either halfword or fullword size (but not both) to be specified. If no specific assignment is made, 12 halfword parameters are assumed.

fields are needed, the flag attached to the symbol for the GENERATE block, shown in Fig. 9-1, is dropped.

9-4
Succession of Events

The program maintains records of when each transaction in the system is due to move. It proceeds by completing all movements that are scheduled for execution at a particular instant of time and that can logically be performed. Where there is more than one transaction due to move, the program processes transactions in the order of their priority class, and on a first-come, first-served basis within priority class.

Normally, a transaction spends no time at a block other than at an ADVANCE block. Once the program has begun moving a transaction, therefore, it continues to move the transaction through the block diagram until one of several circumstances arises. The transaction may enter an ADVANCE block with a non-zero action time, in which case, the program will turn its attention to other transactions in the system and return to that transaction when the action time has been expended. Secondly, the conditions in the system may be such that the action the transaction is attempting to execute by entering a block cannot be performed at the current time. The transaction is said to be blocked and it remains at the block it last entered. The program will automatically detect when the blocking condition has been removed and will start to move the transaction again at that time. A third possibility is that the transaction enters a TERMINATE block, in which case it is removed from the simulation.[4] A fourth possibility is that a transaction may be put on a chain. This concept will be explained in the next chapter.

When the program has moved one transaction as far as it can go, it turns its attention to any other transactions due to move at the same time instant. If all such movements are complete, the program advances the clock to the time of the next most imminent event and repeats the process of executing events.

9-5
Choice of Paths

The TRANSFER block allows some location other than the next sequential location to be selected. The choice is normally between two blocks referred to as next blocks *A* and *B* (the terms *exits 1* and *2* are also used). The method used for

[4]The only other block type that can destroy a transaction is the ASSEMBLE block, which waits for a specified number of transaction copies that were created at a SPLIT block to arrive before allowing one copy to proceed while the rest are destroyed. See footnote 2, this chapter.

choosing is indicated by a *selection factor* in field A of the TRANSFER block. It can be set to indicate one of nine choices. Next blocks *A* and *B* are placed in fields B and C, respectively. If no choice is to be made, the selection factor is left blank. An unconditional transfer is then made to next block *A*.

Of the other modes of choice, only two will be described here. Others will be described in Sec. 10-5 (see also Table 10-2). A random choice can be made by setting the selection factor, *S*, to a three-digit decimal fraction. The probability of going to next block *A* is then $1 - S$, and to the next block *B* it is *S*. A conditional mode, indicated by setting field A to BOTH, allows a transaction to select an alternate path depending upon existing conditions. The transaction goes to next block *A* if this move is possible, and to next block *B* if it is not. If both moves are impossible, the transaction waits for the first to become possible, giving preference to *A* in the event of simultaneity.

9-6

Simulation of a Manufacturing Shop

To illustrate the features of the program described so far, consider the following simple example. A machine tool in a manufacturing shop is turning out parts at the rate of one every 5 minutes. As they are finished, the parts go to an inspector, who takes 4 ± 3 minutes to examine each one and rejects about 10% of the parts. Each part will be represented by one transaction, and the time unit selected for the problem will be 1 minute.

A block diagram representing the system is shown in Fig. 9-2. The usual

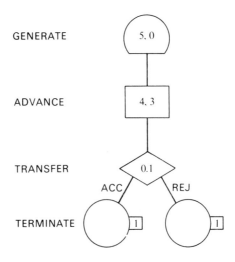

Figure 9-2. Manufacturing shop - model 1.

convention used in drawing blocks is to place the block location (where needed) at the top of the block; the action time is indicated in the center in the form $T = a, b$, where a is the mean and b is the modifier; and the selection factor is placed at the bottom of the block.

A GENERATE block is used to represent the output of the machine by creating one transaction every five units of time. An ADVANCE block with a mean of 4 and modifier of 3 is used to represent inspection. The time spent on inspection will therefore be any one of the values 1, 2, 3, 4, 5, 6, or 7, with equal probability given to each value. Upon completion of the inspection, transactions go to a TRANSFER block with a selection factor of 0.1, so that 90% of the parts go to the next location (exit 1) called ACC, to represent accepted parts and 10% go to another location (exit 2) called REJ to represent rejects. Since there is no further interest in following the history of the parts in this simulation, both locations reached from the TRANSFER block are TERMINATE blocks.

The problem can be coded in a fixed format, as shown in Fig. 9-3. Column 1 is only used for a comment card. An * in column 1 results in the statement being printed in the output only. A field from columns 2 to 6 contains the location of the block where it is necessary for it to be specified. The GPSS program will automatically assign sequential location numbers as it reads the statements, so it is not

LOCATION	OPERATION	A,B,C,D,E,F
*		MANUFACTURING SHOP – MODEL 1
*		
	GENERATE	5 CREATE PARTS
	ADVANCE	4,,3 INSPECT
	TRANSFER	.1,,ACC,REJ SELECT REJECTS
ACC	TERMINATE	1 ACCEPTED PARTS
REJ	TERMINATE	1 REJECTED PARTS
	START	1000

Figure 9-3. Coding of model 1.

usually necessary for the user to assign locations. The TRANSFER block, however, will need to make reference to the TERMINATE blocks to which it sends transactions, so these blocks have been given symbolic location names ACC and REJ.

The second section of the coding, from columns 8 to 18, contains the block type name, which must begin in column 8. Beginning at column 19, a series of fields may be present, each separated by commas and having no embedded blanks. Anything following the first blank is treated as a comment. The meaning of the fields depends upon the block type. Table 9-1 summarizes the information for all the block types that will be described. The table includes references to some block types that will be described in the next chapter.

The program also accepts input in a free format, in which the location field, if used, begins in column 1; but none of the other fields has a fixed starting position. Instead a single blank marks the transition from the location field to the operation field, and from the operation field to the operands. If no location is specified, an initial blank is used. The free format is very convenient when entering input from a terminal, since it minimizes the number of characters that have to be entered. The fixed format, however, is much easier to read, therefore it will be used for the examples given here.

Coding for the GENERATE and ADVANCE blocks was explained in Sec. 9-3. For the TRANSFER block, the first field is the selection factor, and the B and C fields are exits 1 and 2, respectively. In this case, exit 1 is the next sequential block, and it would be permissible to omit the name ACC in both the TRANSFER field B and the location field of the first TERMINATE block. Both commas must, however, be punched to show that field B is missing, so that the TRANSFER block would then be coded: TRANSFER .1,,REJ. It should also be noted that when a TRANSFER block is used in an unconditional mode, so that field A is not specified, a comma must still be included to indicate the field. Thus an unconditional transfer to REJ would be coded as follows, TRANSFER ,REJ.

The program runs until a certain count is reached as a result of transactions terminating. Field A of the TERMINATE block carries a number indicating by how much the termination count should be incremented when a transaction terminates at that block. The number must be positive and it can be zero, but there must be at least one TERMINATE block that has a non-zero field A. In this case, both TERMINATE blocks have 1; so the terminating counter will add up the total number of transactions that terminate, in other words, the total number of both good and bad parts inspected.

The last line shown in Fig. 9-3 is for a control statement called START, which indicates the end of the problem definition and contains, in field A, the value the terminating counter is to reach to end the simulation. In this case the START statement is set to stop the simulation at a count of 1,000. Upon reading a START statement, the program begins executing the simulation.

When the simulation is completed, the program automatically prints an output report, in a prearranged format, unless it has been instructed otherwise. The output automatically generated for the problem just programmed is shown in Fig. 9-4.

```
BLOCK                                                              STATEMENT
NUMBER  *LOC    OPERATION  A,B,C,D,E,F,G,H,I          COMMENTS     NUMBER
                SIMULATE                                           1
        *                                                          2
        * MANUFACTURING SHOP - MODEL 1                             3
        *                                                          4
1               GENERATE   5                CREATE PARTS           5
2               ADVANCE    4,3               INSPECT                6
3               TRANSFER   .1,ACC,REJ        SELECT REJECTS         7
4       ACC     TERMINATE  1                 ACCEPTED PARTS         8
5       REJ     TERMINATE  1                 REJECTED PARTS         9
        *                                                          10
                START      1000              RUN 1000 PARTS         11

                            CROSS-REFERENCE
                            BLOCKS
SYMBOL          NUMBER          REFERENCES

ACC                4               7
REJ                5               7

        SIMULATE
*
* MANUFACTURING SHOP - MODEL 1
*
1       GENERATE   5
2       ADVANCE    4,3
3       TRANSFER   .100,4,5
4       TERMINATE  1
5       TERMINATE  1
*
        START      1000

RELATIVE CLOCK          5005  ABSOLUTE CLOCK        5005
BLOCK COUNTS
BLOCK CURRENT       TOTAL
  1       0         1001
  2       1         1001
  3       0         1000
  4       0          888
  5       0          112
```

Figure 9-4. Output for model 1.

The problem input is printed first, with the locations assigned by the program listed to the left, and a sequential statement number on the right. A table of symbolic locations is then given showing the locations assigned to each symbol. To help the user, the number of the statements that refer to that symbol are also given. A listing of the assembled problem then follows with all symbolic references replaced by numbers.

The first line of output following the listings gives the time at which the simulation stopped. The meaning of the absolute and relative times, which in this case have the same value, will be explained in Sec. 9-10. The time is followed by a listing of block counts. Two numbers are shown for each block of the model. On the left is a count of how many transactions were in the block at the time the simulation stopped, and on the right is a figure showing the total number of transactions that entered the block during the simulation. The results show that the total counts at blocks 4 and 5 were 888 and 112 respectively, showing that of the 1,000 parts inspected, 88.8% were accepted and 11.2% were rejected. In the present example, this is the only output given. Other types of output will be described as more details of the program are explained.

The symbol tables and the assembled form of the input can be suppressed, and usually are, once a problem has been correctly defined. It is also possible to suppress the entire automatic output, and substitute a report defined by the user with an output editor. This allows any item of information that normally appears in the automatic output to be selected, together with any of the items of data called SNA's, which will be described in Chap. 10. These items of data can be arranged in any format, along with comments and headings supplied by the user. It is also possible to produce graphic output in the form of printed graphs.

For the examples used here, the standard output will be used, with the symbol tables and assembled forms suppressed. To save space, the statement numbers will also be dropped.

9-7

Facilities and Storages

Associated with the system being simulated are many permanent entities, such as items of equipment, which operate on the transactions. Two types of permanent entities are defined in GPSS to represent system equipment.

A *facility* is defined as an entity that can be engaged by a single transaction at a time. A *storage* is defined as an entity that can be occupied by many transactions at a time, up to some predetermined limit. A transaction controlling a facility, however, can be interrupted or preempted by another transaction. In addition, both facilities and storages can be made unavailable, as would occur if the equipment they represent breaks down, and can be made available again, as occurs when a repair has been made.

There can be many instances of each type of entity to a limit set by the program (usually 300). Individual entities are identified by number, a separate number sequence being used for each type. The number 0 for these, and all other GPSS entities, is illegal. The user may assign the numbers, in any order, or he may use symbolic names and let the assembly program assign the numbers. Some examples of how the system entities might be interpreted in different systems are:

Type of System	Transaction	Facility	Storage
Communications	Message	Switch	Trunk
Transportation	Car	Toll booth	Road
Data processing	Record	Key punch	Computer memory

A trunk means a cable, consisting of many wires, which can carry several messages simultaneously and is therefore represented by a storage. A switch in this case is assumed to pass only one message at a time and is represented by a facility.

Figure 9-1 shows four block types, SEIZE, RELEASE, ENTER, and LEAVE, concerned with using facilities and storages. Field A in each case indicates which facility or storage is intended, and the choice is usually marked in the flag attached to the symbols of the blocks. The SEIZE block allows a transaction to engage a facility if it is available. The RELEASE block allows the transaction to disengage the facility. In an analogous manner, an ENTER block allows a transaction to occupy space in a storage, if it is available, and the LEAVE block allows it to give up the space. If the fields B of the ENTER and LEAVE blocks are blank, the storage contents are changed by 1. If there is a number (≥ 1), then the contents change by that value. Any number of blocks may be placed between the points at which a facility is seized and released to simulate the actions that would be taken while a transaction has control of a facility. Similar arrangements apply for making use of storages.

To illustrate the use of these block types, consider again the manufacturing shop discussed before. Since the average inspection time is 4 minutes and the average generation rate of parts is one every 5 minutes, there will normally be only one part inspected at a time. Occasionally, however, a new part can arrive before the previous part has completed its inspection. With the diagram of Fig. 9-2, this situation will result in more than one transaction being at the ADVANCE block at one time, and the simulation must be interpreted as meaning that more than one inspector is available.

Assuming that there is only one inspector, it is necessary to represent the inspector by a facility, to simulate the fact that only one part at a time can be inspected. The block diagram will then appear as shown in Fig. 9-5. A SEIZE block and a RELEASE block have been added to simulate the engaging and disengaging of the inspector.

Coding and results for this model are given in Fig. 9-6. A line of output is given for each facility, showing how many times it was seized, the average utilization made of the facility, and the average time transactions held the facility. In this case, the results show that the inspector was busy for 73.5% of his time.

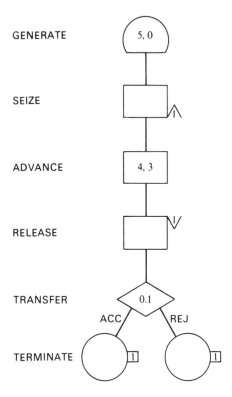

GENERATE

SEIZE

ADVANCE

RELEASE

TRANSFER

TERMINATE

Figure 9-5. Manufacturing shop - model 2.

```
         * MANUFACTURING SHOP - MODEL 2
         *
1                GENERATE    5              CREATE PARTS
2                SEIZE       1              GET INSPECTOR
3                ADVANCE     4,3            INSPECT
4                RELEASE     1              FREE INSPECTOR
5                TRANSFER    .1,ACC,REJ     SELECT REJECTS
6        ACC     TERMINATE   1              ACCEPTED PARTS
7        REJ     TERMINATE   1              REJECTED PARTS
         *
                 START       1000           RUN 1000 PARTS
```

```
RELATIVE CLOCK          5435  ABSOLUTE CLOCK          5435
BLOCK COUNTS
BLOCK CURRENT    TOTAL
    1        1        1001
    2        0        1000
    3        0        1000
    4        0        1000
    5        0        1000
    6        0         900
    7        0         100
```

```
         ************************************************
         *                                              *
         *                  FACILITIES                  *
         *                                              *
         ************************************************
```

FACILITY	NUMBER ENTRIES	AVERAGE TIME/TRAN	-AVERAGE UTILIZATION DURING-				PERCENT AVAILABILITY
			TOTAL TIME	AVAIL. TIME	UNAVAIL. TIME	CURRENT STATUS	
1	1000	3.995	.735				100.0

Figure 9-6. Coding and results for model 2.

If more than one inspector is available, they can be represented as a group by a storage with a capacity equal to the number of inspectors. The SEIZE and RELEASE blocks of Fig. 9-5 would then be replaced by an ENTER and a LEAVE block, respectively. Suppose, for example, the inspection time were three times as long as before; three inspectors would be justified and the model would be coded as listed in Fig. 9-7. Note that a control statement called STORAGE has been added. In the location field, it identifies the storage, and the A field carries the capacity of the storage ($\leq$ 2, 147, 483, 647). Results for this model are also shown in Fig. 9-7. A line of output appears for each storage, giving the information indicated in the headings. It shows, for example, that, on the average, 2.256 of the inspectors were busy at any time.

The case of a single inspector, shown in Fig. 9-5, could have been modeled by using a storage with capacity 1 instead of a facility. An important logical difference between the two possible representations is that, for a facility, only the transaction that seized the facility can release it, whereas, for a storage, entering and leaving can be separate actions carried out independently by different transactions. No use was made of this property in these models, but the difference is of importance in many models.

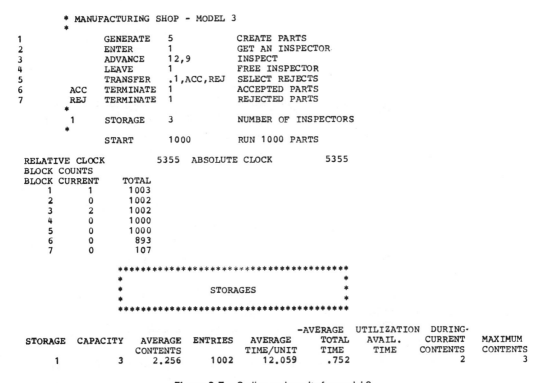

```
    * MANUFACTURING SHOP - MODEL 3
    *
1           GENERATE    5           CREATE PARTS
2           ENTER       1           GET AN INSPECTOR
3           ADVANCE     12,9        INSPECT
4           LEAVE       1           FREE INSPECTOR
5           TRANSFER    .1,ACC,REJ  SELECT REJECTS
6     ACC   TERMINATE   1           ACCEPTED PARTS
7     REJ   TERMINATE   1           REJECTED PARTS
    *
    1       STORAGE     3           NUMBER OF INSPECTORS
    *
            START       1000        RUN 1000 PARTS

RELATIVE CLOCK        5355  ABSOLUTE CLOCK          5355
BLOCK COUNTS
BLOCK CURRENT     TOTAL
    1       1         1003
    2       0         1002
    3       2         1002
    4       0         1000
    5       0         1000
    6       0          893
    7       0          107
```

```
**************************************
*                                    *
*              STORAGES               *
*                                    *
**************************************
```

					-AVERAGE	UTILIZATION	DURING-		
STORAGE	CAPACITY	AVERAGE CONTENTS	ENTRIES	AVERAGE TIME/UNIT	TOTAL TIME	AVAIL. TIME	CURRENT CONTENTS	MAXIMUM CONTENTS
1	3	2.256	1002	12.059	.752		2	3

Figure 9-7. Coding and results for model 3.

9-8

Gathering Statistics

Certain block types in GPSS are constructed for the purpose of gathering statistics about the system performance, rather than of representing system actions. The QUEUE, DEPART, MARK, and TABULATE blocks shown in Fig. 9-1 serve this purpose. They introduce two other entities of the GPSS program, *queues* and *tables*. As with facilities and storages, there can be many such entities up to a prescribed limit (usually, 300 for queues and 100 for tables) and they are individually identified by a number or symbolic name.

When the conditions for advancing a transaction are not satisfied, several transactions may be kept waiting at a block. When the conditions are favorable, they are allowed to move on, according to priority and usually by a first-in, first-out rule. No information about the queue of transactions is gathered, however, unless they have been entered into a queue entity. The QUEUE block increases and the DEPART block decreases the queue numbered in field A. If field B is blank, the change is a unit change; otherwise the value of field B (≥ 1) is used. The program measures the average and maximum queue lengths and, if required, the distribution of time spent on the queue.

It is also desirable to measure the length of time taken by transactions to move through the system or parts of the system, and this can be done with the MARK and TABULATE blocks. Each of these block types notes the time a transaction arrives at the block. The MARK block simply notes the time of arrival on the transaction. (If field A is blank, a special word is used. With nPx in field A, the *n*th parameter of type *x* is used.)[5] The TABULATE block subtracts the time noted by a MARK block from the time of arrival at the TABULATE block. The time, referred to as *transit time*, is entered in a table whose number or name is indicated in field A of the TABULATE block. If the transaction entering a TABULATE block has not passed through a MARK block, the transit time is derived by using as a base the time at which the transaction was created. In effect, the transit time of a transaction can be regarded as the time the transaction has been in the system, and the action of the MARK block is to reset the transit time to zero.

To illustrate the use of these blocks, we return to the model of a manufacturing shop having three inspectors. The properties of the GENERATE block are such that, if a transaction is unable to leave the block at the time it is created, no further creations are made until the block is cleared. The situation simulated in Fig. 9-4 therefore assumes that if one part arrives when all inspectors are busy, the machining of further parts stops until the machine is cleared.

---

[5]In GPSS/360, only the parameter number is given. With parameters, more than one mark time can be carried, allowing transit times for overlapping periods to be carried.

A more realistic situation is that parts will accumulate on the inspectors' work bench if inspection does not finish quickly enough. In that case, it would be of interest to measure the size of the queue of work that occurs. This can be done as shown in Fig. 9-8, which expands upon model 3. A QUEUE block using queue number 1 is placed immediately before the ENTER block, and a DEPART block is placed immediately after the ENTER block to remove the part from the queue when

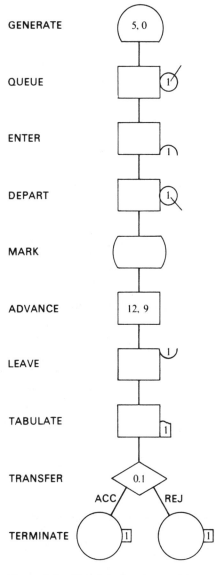

Figure 9-8. Manufacturing shop - model 4.

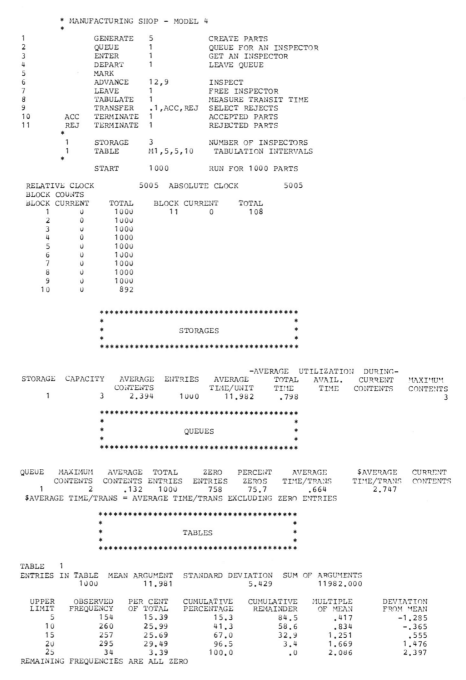

```
        * MANUFACTURING SHOP - MODEL 4
        *
1               GENERATE   5              CREATE PARTS
2               QUEUE      1              QUEUE FOR AN INSPECTOR
3               ENTER      1              GET AN INSPECTOR
4               DEPART     1              LEAVE QUEUE
5               MARK
6               ADVANCE    12,9           INSPECT
7               LEAVE      1              FREE INSPECTOR
8               TABULATE   1              MEASURE TRANSIT TIME
9               TRANSFER   .1,ACC,REJ     SELECT REJECTS
10      ACC     TERMINATE  1              ACCEPTED PARTS
11      REJ     TERMINATE  1              REJECTED PARTS
        *
        1       STORAGE    3              NUMBER OF INSPECTORS
        1       TABLE      M1,5,5,10      TABULATION INTERVALS
        *
                START      1000           RUN FOR 1000 PARTS

RELATIVE CLOCK        5005  ABSOLUTE CLOCK        5005
BLOCK COUNTS
BLOCK CURRENT     TOTAL    BLOCK CURRENT    TOTAL
    1       0     1000      11      0        108
    2       0     1000
    3       0     1000
    4       0     1000
    5       0     1000
    6       0     1000
    7       0     1000
    8       0     1000
    9       0     1000
   10       0      892
```

```
*****************************************
*                                       *
*              STORAGES                 *
*                                       *
*****************************************
```

```
                                            -AVERAGE UTILIZATION DURING-
STORAGE CAPACITY  AVERAGE  ENTRIES  AVERAGE   TOTAL AVAIL.  CURRENT   MAXIMUM
                  CONTENTS          TIME/UNIT  TIME  TIME    CONTENTS  CONTENTS
   1       3      2.394    1000     11.982    .798                          3
```

```
*****************************************
*                                       *
*               QUEUES                  *
*                                       *
*****************************************
```

```
QUEUE  MAXIMUM  AVERAGE  TOTAL   ZERO    PERCENT  AVERAGE     $AVERAGE   CURRENT
       CONTENTS CONTENTS ENTRIES ENTRIES ZEROS    TIME/TRANS  TIME/TRANS CONTENTS
   1      2      .132    1000    758     75.7     .664        2.747
$AVERAGE TIME/TRANS = AVERAGE TIME/TRANS EXCLUDING ZERO ENTRIES
```

```
*****************************************
*                                       *
*               TABLES                  *
*                                       *
*****************************************
```

```
TABLE   1
ENTRIES IN TABLE  MEAN ARGUMENT  STANDARD DEVIATION  SUM OF ARGUMENTS
        1000          11.981           5.429           11982.000

UPPER   OBSERVED   PER CENT   CUMULATIVE   CUMULATIVE   MULTIPLE   DEVIATION
LIMIT   FREQUENCY  OF TOTAL   PERCENTAGE   REMAINDER    OF MEAN    FROM MEAN
   5       154      15.39       15.3         84.5        .417      -1.285
  10       260      25.99       41.3         58.6        .834       -.365
  15       257      25.69       67.0         32.9       1.251        .555
  20       295      29.49       96.5          3.4       1.669       1.476
  25        34       3.39      100.0          .0       2.086       2.397
REMAINING FREQUENCIES ARE ALL ZERO
```

Figure 9-9. Coding and results for model 4.

inspection begins. Any transaction that does not have to wait for an inspector to become available will move through the queue without delay. Those that must wait will do so in the QUEUE block,[6] and the program will automatically measure the length of stay in the queue. Figure 9-8 also shows a MARK and a TABULATE block placed so that they will measure how long the parts take to be inspected, excluding their waiting time in the queue. If the MARK block is omitted, the tabulated time is the time since the transactions first entered the system.

Coding for this model, together with results, is shown in Fig. 9-9. Note that another control statement, called TABLE, has been introduced. In the location field, it identifies a table; and, in the A field, it indicates the quantity to be tabulated. In the B, C, and D fields of the TABLE statement are the lower limit of the table, the tabulation interval size, and the number of intervals, respectively. The definitions of these terms are the same as those introduced in Sec. 8-10 and illustrated in Fig. 8-13. The symbol M1 in field A of the TABLE card indicates transit time tabulation.

It will be seen in Fig. 9-9 that a line of output occurs for each queue, and a table is printed in accordance with the specifications of the TABLE statement. The TABULATE blocks can be used to gather a variety of other statistics, as will be explained in Sec. 10-2.

9-9

Conditional Transfers

As a last example in this chapter, we illustrate the use of both the conditional and unconditional transfer modes of the TRANSFER block. Consider again the case of three inspectors but suppose that the manufactured parts are put on a conveyor, which carries the parts past inspectors, placed at intervals along the conveyor. It takes 2 minutes for a part to reach the first inspector; if he is free at the time the part arrives, he takes it for inspection. If he is busy at that time, the part takes a further 2 minutes to reach the second inspector, who will take the part if he is not busy. Parts that pass the second inspector may get picked up by the third inspector, who is a further 2 minutes along the conveyor belt; otherwise, they are lost. To keep the model small, only the transit time of the parts will be recorded and the possibility of the inspectors rejecting parts will be ignored.

A block diagram for the system is shown in Fig. 9-10. The movement of parts along the conveyor is represented by ADVANCE blocks, each with an action time of 2 minutes. As a transaction leaves an ADVANCE block, it tests whether an inspector is available by entering a TRANSFER block with a selection factor set to

---

[6]The transaction does not have to stay in the QUEUE block in order to stay in the queue.

BOTH. Exit 1 of each of these TRANSFER blocks leads to a SEIZE block representing one of the inspectors. If the inspector is free at the time the transaction
enters, the transaction will leave by way of exit 1 to take over the facility. If the
inspector is busy, the transaction will pass to the next stage of processing by way of
exit 2.

When parts finish inspection, the transactions go to a single TABULATE block
where the transit time is recorded. Because of the rule by which transactions

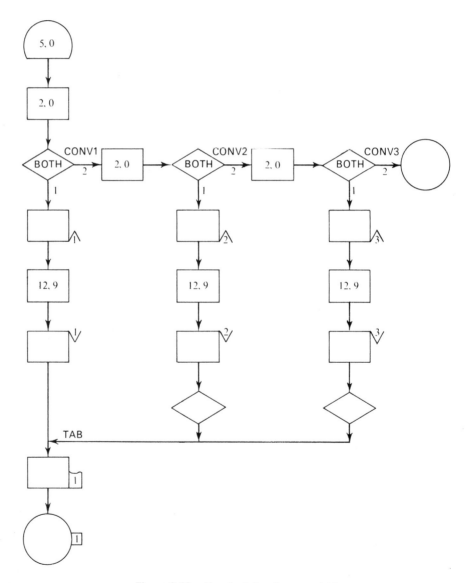

Figure 9-10. Manufacturing shop - model 5.

normally pass from one block to the next higher numbered block, only one of the three RELEASE blocks that complete the inspection phase is able to pass the transactions directly to the TABULATE block. The others pass the transactions to TRANSFER blocks that send the transactions unconditionally to the TABULATE block. It would be possible to have three separate TABULATE blocks at the exit of each RELEASE block, making the unconditional TRANSFER blocks unnecessary. Each TABULATE block can refer to the same table number, so that the two alternative ways of drawing the block diagram will lead to exactly the same statistics. This principle—allowing the same GPSS entity to be referred to in more than one place —applies to all entities.

Coding and results for the problem are shown in Fig. 9-11. An explanation of the extra START statement shown in the coding will be given in the next section. Note how the rule for normally moving transactions to successively numbered blocks causes the coding to be broken into several parts. Breaks are caused by both the unconditional TRANSFER and the conditional TRANSFER blocks. A way of simplifying block diagrams, such as this one, that have repetitive segments will be discussed in Sec. 10-8.

9-10

Program Control Statements

The first statement of the GPSS input is a control statement with the word SIMULATE in the operation field. Without this statement, the problem will be assembled but not executed.

It is desirable to be able to stop and restart a simulation run and also repeat a simulation run with changes to some of the values in the model. When a GPSS simulation run is finished, therefore, the program does not immediately destroy the model. Instead, it looks for more input following the START statement, keeping the model exactly as it was at the completion of the run. Input following the START statement can change the model. For example, a storage capacity could be redefined by inserting a STORAGE statement that refers to the number of a previously defined storage and gives the new value. The model could also be modified by changing existing blocks or adding new blocks. When all such modifications have been read and the model is ready for rerunning another START statement will instruct the program to begin simulation again.

Certain control statements can be included between START statements to set the conditions under which the rerun is made. A statement with the single word RESET in the operation field will wipe out all the statistics gathered so far, but will leave the system loaded with transactions. The purpose is to gather statistics in the second run with the initial build-up period excluded. The output of the first run is

```
* MANUFACTURING SHOP - MODEL 5
*
1             GENERATE   5              CREATE PARTS
2             ADVANCE    2              PLACE ON CONVEYOR
3             TRANSFER   BOTH,,CONV1    MOVE TO FIRST INSPECTOR
4             SEIZE      1              GET FIRST INSPECTOR
5             ADVANCE    12,9           INSPECT
6             RELEASE    1              FREE INSPECTOR
7      TAB    TABULATE   1              MEASURE TRANSIT TIME
8             TERMINATE  1
       *
9      CONV1  ADVANCE    2              PLACE ON CONVEYOR
10            TRANSFER   BOTH,,CONV2    MOVE TO SECOND INSPECTOR
11            SEIZE      2              GET SECOND INSPECTOR
12            ADVANCE    12,9           INSPECT
13            RELEASE    2              FREE INSPECTOR
14            TRANSFER   ,TAB
       *
15     CONV2  ADVANCE    2              PLACE ON CONVEYOR
16            TRANSFER   BOTH,,CONV2    MOVE TO THIRD INSPECTOR
17            SEIZE      3              GET THIRD INSPECTOR
18            ADVANCE    12,9           INSPECT
19            RELEASE    3              FREE INSPECTOR
20            TRANSFER   ,TAB
       *
21     CONV3  TERMINATE
       *
     1        TABLE      M1,5,5,10      TABULATION INTERVALS
       *
              START      10,NP          INITIALIZE WITH TEN PARTS
              RESET
              START      1000           MAIN RUN
```

RELATIVE CLOCK   5588   ABSOLUTE CLOCK   5660
BLOCK COUNTS

BLOCK	CURRENT	TOTAL	BLOCK	CURRENT	TOTAL	BLOCK	CURRENT	TOTAL
1	0	1117	11	0	343	21	0	118
2	0	1117	12	0	343			
3	0	1118	13	0	344			
4	0	406	14	0	344			
5	1	406	15	1	369			
6	0	406	16	0	368			
7	0	1000	17	0	250			
8	0	1000	18	0	250			
9	0	712	19	0	250			
10	0	712	20	0	250			

```
*****************************************
*                                       *
*              FACILITIES               *
*                                       *
*****************************************
```

			-AVERAGE	UTILIZATION	DURING-		
FACILITY	NUMBER ENTRIES	AVERAGE TIME/TRAN	TOTAL TIME	AVAIL. TIME	UNAVAIL. TIME	CURRENT STATUS	PERCENT AVAILABILITY
1	407	11.885	.865				100.0
2	344	11.956	.736				100.0
3	250	12.244	.547				100.0

```
*****************************************
*                                       *
*                TABLES                 *
*                                       *
*****************************************
```

TABLE   1
ENTRIES IN TABLE   MEAN ARGUMENT   STANDARD DEVIATION   SUM OF ARGUMENTS
       1000   15.693   5.757   15694.000

UPPER LIMIT	OBSERVED FREQUENCY	PER CENT OF TOTAL	CUMULATIVE PERCENTAGE	CUMULATIVE REMAINDER	MULTIPLE OF MEAN	DEVIATION FROM MEAN
5	27	2.69	2.6	97.2	.318	-1.857
10	202	20.19	22.8	77.1	.637	-.988
15	269	26.89	49.7	50.2	.955	-.120
20	248	24.79	74.5	25.4	1.274	.747
25	228	22.79	97.3	2.6	1.592	1.616
30	26	2.59	100.0	.0	1.911	2.484

REMAINING FREQUENCIES ARE ALL ZERO

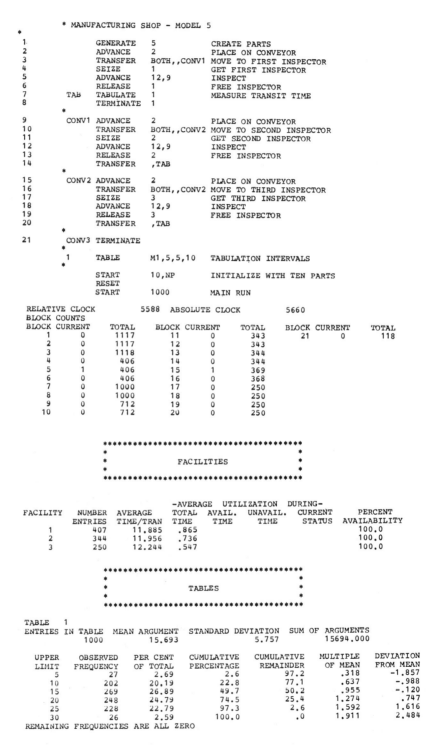

Figure 9-11. Coding and results for model 5.

not then of interest and is usually suppressed by putting the letters NP in the B field of the START statement. This procedure has been followed in the simulation shown in Fig. 9-11. One START statement has run the model for a total of 10 completed transactions, and the output of this run has been suppressed. A RESET statement has wiped out the statistics gathered in that run; the second START statement has restarted the simulation from the point at which the tenth transaction finished and has continued for a further 1,000 transactions. The results shown are the statistics gathered during the second run.

The RESET statement also sets to zero a relative clock. When the run is completed, the output marked ABSOLUTE CLOCK gives the time since the run began; the output marked RELATIVE CLOCK gives the time since the last RESET statement. If the RESET statement is not used, the two times are the same.

Another control statement, CLEAR, would not only wipe out the statistics of the preceding run, but would also wipe out the transactions in the system; so that, the rerun does in fact start the simulation from the beginning. This procedure would be followed when the objective is to repeat the problem with a change in some value. Although the CLEAR statement returns the model to its initial state, it does not reset the random number generator seeds. The following sequence of statements

START

CLEAR

START

would run the same problem twice but the second run would use a different set of random numbers. When there are stochastic processes in the model, the second run will be a second sample of the possible outputs that could arise.

As well as controlling different runs of the same model, GPSS allows any number of different problems to be run successively with a single loading of the program. A control statement with the word JOB in the operation field instructs the program to wipe out the entire model preceding the statement and proceed with the following problem. This can be repeated as many times as desired. All five models described in this chapter were run at one time by placing a JOB statement between the runs. A JOB statement is not needed in front of the first problem. Unlike the CLEAR statement, the JOB statement does reset the random number seeds so that each problem finds the program in exactly the same form as does the first problem.

Since the program always anticipates further input when it completes a run, a control statement with the word END in the operation field must be placed at the end of all problems, even if there is only one, to terminate all simulation. The

SIMULATE statement appears only once, at the beginning of the input deck, even if there are multiple jobs.

Exercises

Give GPSS block diagrams and write programs for the following problems:

9-1   In the manufacturing shop model of Fig. 9-2, suppose that parts rejected by the inspector are sent back for further work. Reworking takes 15 ± 3 minutes and does not involve the machine that originally made the parts. After correction, the parts are resubmitted for inspection. Simulate for 1,000 parts to be completed.

9-2   Parts are being made at the rate of one every 6 minutes. They are of two types, A and B, and are mixed randomly, with about 10% being type B. A separate inspector is assigned to examine each type of part. The inspection of A parts takes 4 ± 2 minutes and B parts take 20 ± 10 minutes. Both inspectors reject about 10% of the parts they inspect. Simulate for a total of 1,000 type A parts accepted.

9-3   Workers come to a supply store at the rate of one every 5 ± 2 minutes. Their requisitions are processed by one of two clerks who take 8 ± 4 minutes for each requisition. The requisitions are then passed to a single storekeeper who fills them one at a time, taking 4 ± 3 minutes for each request. Simulate the queue of workers and measure the distribution of time taken for 1,000 requisitions to be filled.

9-4   People arrive at an exhibition at the rate of one every 3 ± 2 minutes. There are four galleries. All visitors go to gallery A. Eighty percent then go to gallery B, the remainder go to gallery C. All visitors to gallery C move on to gallery D and then leave. None of the visitors to gallery B goes to gallery C; however, about 10% of them do go to gallery D before leaving. Tabulate the distribution of the time it takes 1,000 visitors to pass through when the average times spent in the galleries are as follows:

$$A, 15 \pm 5; \quad B, 30 \pm 10; \quad C, 20 \pm 10; \quad D, 15 \pm 5$$

9-5   People arrive at the rate of one every 10 ± 5 minutes to use a single telephone. If the telephone is busy, 50% of the people come back 5 minutes later to try again. The rest give up. Assuming a call takes 6 ± 3 minutes, count how many people will have given up by the time 1,000 calls have been completed.

9-6   People arrive at a cafeteria with an inter-arrival time of 10 ± 5 seconds. There are two serving areas, one for hot food and the other for sandwiches.

The hot food area is selected by 80% of the customers and it has 6 servers. The sandwich area has only one server. Hot food takes 1 minute to serve and sandwiches take $\frac{1}{2}$ minute. When they have been served, the customers move into the cafeteria, which has seating capacity for 200 people. The average time to eat a hot meal is 30 ± 10 minutes and for a sandwich, 15 ± 5 minutes. Measure the queues for service and the distribution of time to finish eating, from the time of arrival. Simulate for 1,000 people to finish their meal.

9-7   Ships arrive at a harbor at the rate of one every $1 \pm \frac{1}{2}$ hours. There are six berths to accommodate them. They also need the services of a crane for unloading and there are five cranes. After unloading, 10% of the ships stay to refuel before leaving; the others leave immediately. Ships do not need the cranes for refueling. Simulate the queues for berths and cranes assuming it takes $7\frac{1}{2} \pm 3$ hours to unload and $1 \pm \frac{1}{2}$ hours to refuel. Simulate for 100 ships to clear the harbor.

Bibliography

1   EFRON, R., AND G. GORDON, "A General Purpose Digital Simulator and Examples of its Application: Part 1—Description of the Simulator," *IBM Syst. J.*, III, no. 1 (1964), 21–34.

2   GORDON, GEOFFREY, "A General Purpose Systems Simulation Program," in *Proc. EJCC, Washington, D.C.*, pp. 87–104, New York: Macmillan Publishing Co., Inc., 1961.

3   ———, *The Application of GPSS V to Discrete System Simulation*, Englewood Cliffs, N.J.: Prentice-Hall, Inc., 1975.

4   GOULD, R. L., "GPSS/360 — An Improved General Purpose Simulator," *IBM Syst. J.*, VIII, no. 1 (1969), 16–27.

5   HERSCOVITCH, H., AND T. SCHNEIDER, "GPSS III — An Expanded General Purpose Simulator," *IBM Syst. J.*, IV, no. 3 (1965), 174–183.

6   IBM Corp., *General Purpose Simulation System V, User's Manual* (Form SH 20–0851), White Plains, N.Y.

7   ———, *General Purpose Simulation System V, Reference Card — Block Statement Formats* (Form GX20 1828).

8   ———, *General Purpose Simulation System V Reference Card — Control Statement Formats* (Form GX20 1928).

10

GPSS EXAMPLES

10-1

Priorities and Parameters

The previous chapter introduced the concepts of the GPSS program, described some of the block types, and illustrated the manner in which the program is used. In this chapter, a fuller account of the program will be given together with descriptions of more block types. (See Fig. 9-1 for the block symbols, Table 9-1 for coding information for the blocks, and Table 9-2 for control statements.)

As described so far, transactions have no particular identity; each is treated by a block in the same manner as any other transaction. In fact, transactions have two types of attribute which influence the way they are processed. Each transaction has one of 128 levels of *priority*, indicated by the numbers 0 to 127, with 0 being the lowest priority. At any point in the block diagram, the priority can be set up or down to any of the levels by the PRIORITY block. The block is coded by putting the priority in field A of the block. It is also possible to designate the priority at the time a transaction is created by putting the priority in the E field of the GENE-RATE block creating the transaction. If the field is left blank, the priority is set to 0.

When there is competition between transactions to occupy a block, the service rule is to advance transactions in order of priority and first-in, first-out within priority class. The program has the capability of implementing more complex queuing disciplines through the chain feature described in Sec. 10-9.

A transaction also has parameters, which carry numerical data that can affect the way the transaction is processed by a block. The values are identified by the notation Pxn, where n is the parameter number, and x is the type. Parameters were defined in Sec. 9-3 when describing the formulation of the GENERATE block, since it is then that the parameters associated with a transaction are determined.[1] As stated there, if no declaration is made, the transactions have 12 halfword parameters, and that is the case that will be used in the examples given here. The notation for parameters will, therefore, be PHn.

All parameter values are zero at the time a transaction is created. A value is given to a parameter when a transaction enters an ASSIGN block. The number of the parameter is given in field A of the ASSIGN block. The value the parameter is to take is given in field B; the value can be a specific number or can name any of the SNA's, which are described in the next section, as the source of the value. Field C has one of the letters F, H, B, or L to indicate the type of the parameter. A frequent requirement is to assign a value that is to be used later as an action time. If that is the case, the B and C fields are used as though they were a mean and a modifier in the form of a function. The designation of the type of parameter is then moved to field D.

An ASSIGN block can either add to, subtract from, or replace the value of a parameter. A + or − sign immediately following the parameter number in field A indicates that the assigned value is to be added or subtracted. For replacement, the parameter number only is given.

Usually, parameters record characteristics of the entity represented by the transaction. For example, a transaction representing a message in a communication system might use parameters to represent the message length, the message type, or the destination. A transaction representing a ship might use parameters for carrying the cruising speed, number of passengers, or type of cargo. Another use of parameters is for logical control of the model. A transaction might, for example, be controlling the number of times an operation is to be performed, with parameters carrying the number of times the operation has been performed so far and the total number required.

10-2

Standard Numerical Attributes

Parameters are items of data that represent attributes of the transaction. In addition, there are attributes of the other entities of the system, such as the number of transactions in a storage or the length of a queue, which are made available to

---

[1] A slightly different notation was used in Sec. 9-3. The notation used there is for indicating the number and type of parameter the program is to process. The present notation is used when the program is to use the current value of the parameter as a system variable.

the program user. Collectively, they are called *standard numerical attributes (SNA's)*. Each type of SNA is identified by a one or two letter code and a number. For example, the contents of storage number 5 is denoted by S5 and the length of queue number 15 is Q15. For completeness, parameters are included in the category of SNA's.

Computations can be carried out by defining *variable statements* in which simple SNA's are combined mathematically with the operators $+$, $-$, $\star$, $/$, and @ for addition, subtraction, multiplication, division, and modulo division, that is, deriving the remainder after division. Parentheses, up to five levels, may be included in the definition of a variable statement. A variable statement is numbered and defined by a statement that has the number, or name, in the location field and the word VARIABLE in the operation field. The statement begins in column 19 and continues without blanks up to column 71 if necessary. For example, the following statement:

```
5 VARIABLE    S6+5★(Q12+Q17)
```

defines variable 5 as being the sum of the current contents of storage number 6 plus 5 times the sum of the lengths of queues 12 and 17. A variable statement operates with integers and gives an integer as output. If a floating-point number should occur in its definition, the value of the number will be truncated. When a floating-point calculation is needed, the operation field is changed to the notation FVARIABLE. The definition is otherwise the same as for a variable.

Many uses can be made of SNA's. They provide the inputs to the functions described in the next section, thereby allowing a great variety of functional relationships to be introduced into the model. It will be recalled that the TABLE statement used in conjunction with the TABULATE block to measure transit time carried M1 in field A. Transit time is one of the SNA's and M1 is its symbol; any other SNA can be used in a TABLE statement, allowing the program to collect and tabulate a wide variety of statistics. The current value of any SNA may also be used as the value of almost any field of a block.

In general, the values of the SNA's change as the simulation proceeds. The program does not continuously maintain current value; it computes the value of SNA's at the time they are needed. It is convenient to be able to save values computed at one time for use at some later time, and this can be done by the use of a block type called SAVEVALUE. The block indicates in field A the number of one of many *savevalue locations* and, in field B, gives the SNA to be saved. As with an ASSIGN block, a $+$ or $-$ sign immediately following the savevalue number will result in the SNA value being added to or subtracted from the savevalue contents; otherwise, the value is replaced.

Savevalues occur with the same four formats used for parameters. The C field of the SAVEVALUE block must carry one of the symbols XF, XH, XB, or XL to indicate a fullword, halfword, byte-sized, or floating-point type of savevalue. For example, the following block would save the current contents of storage number 6 in halfword savevalue number 10

SAVEVALUE   10,S6,XH

The program also allows two-dimensional matrices of savevalues to be defined. They also occur in four different formats.

The current contents of a savevalue location n is available as an SNA indicated by Xxn, where x denotes the type of savevalue. A control statement, called INITIAL can set the value of a savevalue at the beginning of the simulation so that savevalues can be used to introduce data and set initial conditions. The notation Xxn is put in field A and the initial value is put in field B. The values of the non-zero save-values are printed at the end of the simulation run, so that savevalues also provide a way of extracting results from the simulation other than through the standard report.

A partial list of the program SNA's is given in Table 10-1. The letter n in Table 10-1 is to be replaced by the number of some entity. The program allows any entity (but not a parameter) to be represented by a symbolic name. If a symbolic name is used, a reference to the entity in an SNA must use the name preceded by a $ sign. For example, a storage called BSKT would be referred to as S$BSKT.

TABLE 10-1 GPSS Standard Numerical Attributes

C1	The current value of clock time.
CHn	The number of transactions on chain n.
Fn	The current status of facility number n. This variable is 1 if the facility is busy and 0 if not.
FNn	The value of function n. (The function value may be computed to have a fractional part but the SNA gives only the integral part, unrounded.)
Kn	The integer n (the notation n may also be used).
M1	The transit time of a transaction.
Nn	The total number of transactions that have entered block n.
Pxn	Parameter number n of a transaction, of type x.
Qn	The length of queue n.
Rn	The space remaining in storage n.
RNn	A computed random number having one of the values 1 through 999 with equal probability. (When the reference is made to provide the input for a function, the value is automatically scaled to the range 0 to 1.) Eight different generators can be referenced by n = 1, 2, . . . , 8.
Sn	The current occupancy of storage n.
Vn	The value of variable statement number n.
Wn	The number of transactions currently at block n.
Xxn	The value of savevalue location n of type x.

10-3

Functions

To introduce functional relationships in a model, a GPSS model can include a number of functions. Each function is defined by giving two or more pairs of numbers that relate an input x to an output y. The function can be in a continuous or discrete mode. In a continuous mode, the program will interpolate linearly between the defined points, allowing the user to approximate a continuous function by a series of straight line segments. In a discontinuous mode, the function is treated as a "staircase function." If x_i and x_{i+1} are two successive points at which the fu iction has been defined, an input value in the range $x_i < x \leq x_{i+1}$ will result in the value y_i being produced. Figure 10-1 illustrates these two modes of using functions.

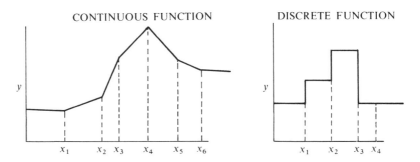

Figure 10-1. Continuous and discrete functions.

Any number of points (> 1) can be used to define a function; and the size of the intervals between successive points can vary freely, except that a function defined to have a random number as input must have values of x in the range 0 to 1. The values of x and y can be fractional and negative. A reference to a value of x below the lowest defined value, x_1, results in the value y_1. Similarly, reference beyond the highest defined value of x produces the value of y at that highest value.

In using a function, any of the SNA's can be defined as being the input. Examples of some of the more commonly used choices of input and their uses are:

(a) By using a uniformly distributed random number **RNn**, as input, any other distribution can be obtained with the technique of Sec. 6-10.

(b) By using clock time **C1**, action times can be made to depend upon clock time, thereby simulating the effects of peak loads.

(c) By using the current contents of a storage, an action time can depend upon the current load on some part of the system.

(d) By using a parameter, an action time can depend on each particular transaction.

The choice of input is made at the time of defining the function. It is *not* selected when the function is referenced. At least two statements are needed to define a function. The first is called a FUNCTION statement and it is immediately followed by one or more *function data* statements. The FUNCTION statement has the following format:

Field	Contents
Location	Function number
Operation	FUNCTION
A	SNA to be used as input
B	Cn for continuous mode
	Dn for discontinuous mode, where n is the number of points to be defined.

The function data statements carry the values of x and y that define the function. Values begin in column 1 with commas between the x and y values and slashes between successive pairs thus: $x_1, y_1/x_2, y_2/, \ldots$. Any number of points may be in one statement as long as the statement does not go beyond column 71. Any number of additional statements may be used but no one pair of $x\,y$ values may be split between two statements. Some examples will be shown in the next section.

10-4

Simulation of a Supermarket

To illustrate the use of functions, parameters, and SNA's, a simulation model will be written for a supermarket that operates in the following manner. Customers of the supermarket are obliged to take a basket before they begin to shop. There is a limited number of baskets and, if no basket is available when they arrive, customers leave without shopping. If they get a basket, customers shop and then check-out at one of five check-out counters. After checking-out, they return the baskets and leave the supermarket. A block diagram of the model is shown in Fig. 10-2. There are four sections, concerned with

(a) Getting a basket
(b) Shopping
(c) Checking-out
(d) Leaving

Each shopper is represented by a transaction and the unit of time is 1 second.

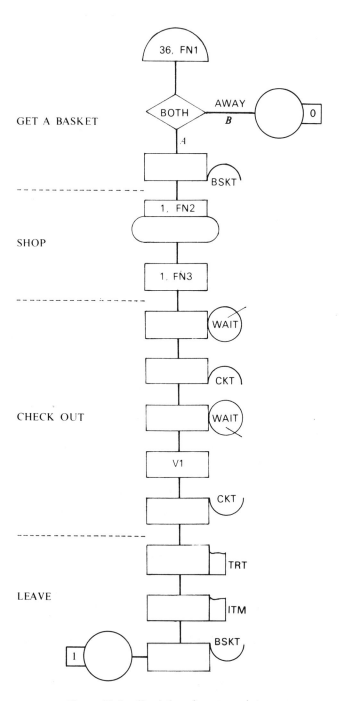

Figure 10-2. Simulation of a supermarket.

A GENERATE block creates the transactions that represent the customers. It is assumed that the arrival pattern can be represented by a Poisson distribution. The process for generating such numbers requires the function $y = -\ln(1 - x)$. It will be recalled that this function was approximated by the numbers of Table 7-4, which is described in Sec. 7-4. Function number 1 of the simulation model uses these data, and controls the inter-arrival time at a GENERATE block with a mean of 36.

To represent the baskets, a storage denoted by BSKT is used with capacity equal to the number of baskets; in this case, there will be 50 baskets. The decision whether a customer is able to get a basket is made at a TRANSFER block immediately following the GENERATE block. The TRANSFER block has a BOTH selection factor and it attempts to send transactions to an ENTER block using the storage BSKT. If a basket is available, the ENTER block will accept the transaction and increase the count of baskets in use by 1. If the storage is full, however, no basket is available, and the TRANSFER block sends the transaction to a TERMINATE block called AWAY. The block count at AWAY will show the number of customers turned away for lack of baskets. Coding for this section is

```
          GENERATE     36,FN1
          TRANSFER     BOTH,,AWAY
          ENTER        BSKT
          ............
AWAY      TERMINATE
```

The simulation will be arranged to select the number of items purchased. A parameter, number 1, is assigned to represent the number of items to be purchased. The number of items is determined by an ASSIGN block which has the SNA called FN2 in field B. The number of items is therefore picked from function number 2 which is a discrete function, defined as follows:

x	y
$0 < x \leq 0.2$	5
$0.2 < x \leq 0.5$	10
$0.5 < x \leq 0.9$	15
$0.9 < x \leq 1.0$	20

The input is RN1, which is the SNA denoting the output of random generator number 1. Since this is the input to a function, the value is automatically scaled to the range 0 to 1. The technique described in Sec. 6-9 is, therefore, being used to produce one of the numbers 5, 10, 15, or 20, with probabilities 0.2, 0.3, 0.4, and 0.1, respectively.

Transactions then pass to an ADVANCE block to represent the shopping time. The time will be made to depend upon the number of shoppers in the supermarket, in accordance with the continuous function shown in Fig. 10-3. The input to the function is the SNA called S$BSKT which is the current contents of the storage representing the baskets. Since we specify that a customer must have a basket, the value of S$BSKT is the number of customers engaged in shopping. The shopping time is coded as function number 3. The ADVANCE block representing the shopping has FN3 in the A field, so the value of the function becomes the action time.

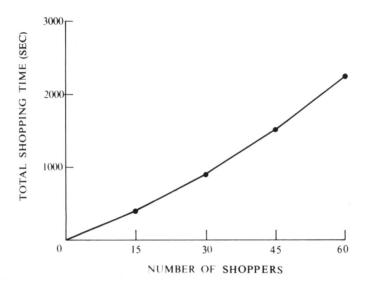

Figure 10-3. Shopping-time function.

When transactions emerge from the ADVANCE block, shopping is completed and they move to the section concerned with checking-out. There are five counters but it is not necessary in this study to distinguish the performance of the individual counters. The counters are therefore regarded as a service unit with five servers. They are represented by a storage, named CKT, that has a capacity of five.

There will be some congestion at the counters, and one of the objectives will be to measure the amount of congestion. The transactions therefore go to a QUEUE block, which enters them in a queue called WAIT. When a counter becomes available, a transaction leaves the QUEUE block for the ENTER block and then immediately goes to a DEPART block to be removed from the queue. Should there be no congestion, a transaction will move straight through the QUEUE, ENTER, and DEPART blocks.

Checking-out will be assumed to require 10 seconds to pay for each item plus 25 seconds for packing. Since parameter 1 is the number of items, the following variable statement will compute the checking-out time:

1   VARIABLE   PH1★10+25

Field A of an ADVANCE block representing the checking-out time is set to V1. As each transaction enters the block, the program computes the action time from variable statement number 1. When checking-out is completed, the transactions go to a LEAVE block to give up the counter space. The coding for the check-out section is

QUEUE	WAIT
ENTER	CKT
DEPART	WAIT
ADVANCE	V1
LEAVE	CKT

Upon completion of check-out, transactions pass to the section concerned with leaving the supermarket. They first go to a TABULATE block where the transit time is tabulated in a table called TRT. Suppose a record is also to be made of the number of items bought by each customer. This is done by going to another TABULATE block which tabulates PH1 in a table called ITM. The tabulation should, of course, reproduce the original distribution of function 2. This step has been inserted to illustrate the use of a TABULATE block for statistics other than transit time.

When tabulation is complete, transactions go to a LEAVE block, naming the storage BSKT, to represent the return of the basket. Finally they go to a TER-MINATE block. Coding for this section is

	TABULATE	TRT
	TABULATE	ITM
	LEAVE	BSKT
	TERMINATE	1
TRT	TABLE	M1,500,500,10
ITM	TABLE	PH1,5,5,4

Note that a 1 appears in the A field of this TERMINATE block while the A field of the other TERMINATE block is blank. The simulation run, therefore, will count only satisfied customers.

Coding for the complete model is shown in Fig. 10-4. Note a restraint on the order in which statements must be given in this example. Because a reference is

```
* SIMULATION OF A SUPERMARKET
*
1       FUNCTION   RN1,C24              FUNCTION FOR I/A INTERVAL
0.0,0.0/0.1,0.104/0.2,0.222/0.3,0.355/0.4,0.509/0.5,0.69
0.6,0.915/0.7,1.2/0.75,1.38/0.8,1.6/0.84,1.83/0.88,2.12
0.9,2.3/0.92,2.52/0.94,2.81/0.95,2.99/0.96,3.2/0.97,3.5
0.98,3.9/0.99,4.6/0.995,5.3/0.998,6.2/0.999,7/0.9997,8
*
        GENERATE   36,FN1               CREATE SHOPPERS
        TRANSFER   BOTH,,AWAY           CHECK FOR AVAILABLE BASKET
        ENTER      BSKT                 GET A BASKET
        ASSIGN     1,FN2,PH             DETERMINE NUMBER OF ITEMS
        ADVANCE    FN3                  SHOP
        QUEUE      WAIT                 WAIT FOR COUNTER SPACE
        ENTER      CKT                  GET COUNTER SPACE
        DEPART     WAIT                 LEAVE WAITING LINE
        ADVANCE    V1                   CHECK-OUT
        LEAVE      CKT                  FREE COUNTER SPACE
        TABULATE   TRT                  TABULATE TRANSIT TIME
        TABULATE   ITM                  TABULATE NUMBER OF ITEMS
        LEAVE      BSKT                 RETURN BASKET
        TERMINATE  1
*
AWAY    TERMINATE                       LOST CUSTOMERS
*
TRT     TABLE      M1,500,500,10        TRANSIT TIME TABLE
ITM     TABLE      PH1,5,5,4            ITEM COUNT TABLE
*
CKT     STORAGE    5                    NUMBER OF COUNTERS
BSKT    STORAGE    50                   NUMBER OF BASKETS
*
2       FUNCTION   RN1,D4               DISTRIBUTION OF ITEMS
.2,5/.5,10/.9,15/1.0,20
*
3       FUNCTION   S$BSKT,C5            SHOPPING TIME
0,0/15,400/30,900/45,1500/60,2250
*
1       VARIABLE   PH1*10+25            CHECK-OUT TIME
*
        START      50,NP                INITIALIZE, SUPPRESS PRINTING
        RESET
        START      1000                 MAIN RUN
```

Figure 10-4. Coding of supermarket simulation.

made to a function by the GENERATE block, the function must be defined ahead of the GENERATE block. There is otherwise no restriction on where a function can appear in the coding.

10-5

Transfer Modes

A further important use of both parameters and functions is in conjunction with the TRANSFER block. Rather than just making a random choice or a conditional selection between two blocks, a TRANSFER block can use the value of a parameter or a function as the location to which it sends a transaction. To use the parameter mode, Px is put in the A field of the TRANSFER block, where x denotes the type, and the parameter number is put in field B. For the function mode, FN is put in field A and the function number is put in field B. If a number is put in field C, it is added to the parameter or function value. Both modes of operation are particularly useful when different categories of transactions must be handled in the same way in one part of the block diagram but, eventually, must be separated.

Continuous or discrete functions using any SNA as input can be employed, but a particular mode of function has been defined to simplify the use of functions for transfers. A *list-mode* function assumes that the input is an integer n, and it returns the nth listed value of the function. Since location numbers are not always known before assembly, it is permissible to use location names as function values; the assembly program will supply the correct numerical value. Suppose, for example, transactions are to be sent to one of four locations called LOCA, LOCB, LOCC, and LOCD, and suppose further that one of the numbers 1, 2, 3, or 4 has been placed in a parameter, say number 3. The following coding will make the transfer:

```
           TRANSFER     FN,1
           . . . . . .
1          FUNCTION     PH3,L4
1,LOCA/2,LOCB/3,LOCC/4,LOCD
```

The characters L4 signify that the function is in the list mode and has four listed values.

Alternatively, an ASSIGN block could have used this function to set a parameter, say number 1, to be the location. Later, the transfer could be effected with the block

```
           TRANSFER     PH,1
```

Table 10-2 summarizes all the modes of the TRANSFER block that have been described.

TABLE 10-2 GPSS Transfer Modes

Mode	Field A	Field B	Field C
Unconditional		Next Block A	
Random	.xxx	Next Block A	Next Block B
Conditional	BOTH	Next Block A	Next Block B
Parameter	Px	Parameter No.	(Increment)
Function	FN	Function No.	(Increment)

10-6

Logic Switches

Two types of entities, facilities and storages, have been introduced to represent equipment. In addition, a third type of entity called *logic switches* is made available to represent two-state conditions in a system. Each switch is either on or off, and a block type called LOGIC is used to change the status of the switch. A transaction entering the block can either set the switch on, reset the switch off, or invert the switch from its current state. Should the switch already be in the desired state, no action is taken. In coding the LOGIC block, the letter S, R, or I indicating set, reset, or invert appears in *column 14* while the switch number, or name, appears in field A. The program keeps no statistics about logic switches. However, it prints the numbers of the switches that are set at the time the run ends.

Examples of how logic switches are used are: a model representing a factory might use a logic switch to indicate whether a machine is in working order or not; in a vehicular traffic system, a logic switch might represent a traffic light; and, in a bank, a logic switch might represent whether a teller's position is open.

10-7

Testing Conditions

It is often desirable to control the flow of transactions according to prevailing conditions in the system. A block type called GATE can be used for this purpose. It can test the condition of any facility, storage, or logic switch in the block diagram. Some of the conditions that can be tested and the symbols used to indicate the selected conditions are

LS	n	Logic switch n set
LR	n	Logic switch n reset
U	n	Facility n in use
NU	n	Facility n not in use
SF	n	Storage n full
SNF	n	Storage n not full
SE	n	Storage n empty
SNE	n	Storage n not empty

A transaction will enter the GATE block if the condition being tested is true. When the condition is not true, there is a choice of action. If an alternative block is specified, the transaction will be sent to the alternative block immediately. If no alternative is specified, the transaction will wait until the tested condition becomes true and then enter the GATE block. It is not necessary for the user to arrange for retesting; the program will automatically recognize when the condition changes and move the transaction at that time. In coding GATE blocks, the condition code begins in *column 13*, while the number or name of the entity to be tested is placed in field A. If an alternative block is specified, it is put in field B.

Another block type called TEST can test a variety of relationships between any two SNA's. Since the SNA's include variable statements, the TEST block is able to perform complex tests of conditions. The relationships that can be tested and the symbols used to represent them are shown below:

G	Greater than
GE	Greater than or equal
E	Equal
NE	Not equal
LE	Less than or equal
L	Less than

The relationship symbol begins in *column 13* and the two SNA's to be related are placed in fields A and B. Thus, the following block will test whether the content of storage 6 is less than the number recorded in halfword savevalue location 12

```
          TEST L        S6,XH12
```

The action taken upon testing the condition is the same as for a GATE block. A transaction will enter the block if the condition is true and will either go to an alternative block if one is specified (in field C) or wait to enter when the condition changes.

10-8

GPSS Model of a Simple Telephone System

To illustrate the use of logic switches, a GPSS model of the telephone system discussed in Chap. 8 will be derived. The system is one in which a series of calls come from a number of telephone lines and the system is to connect the calls by using one of a limited number of links. Only one call can be made to any one line at a time and it is assumed that calls are lost if the called party is busy or no link is available. Each line is represented by a logic switch whose number is the line number. The line is considered busy if the switch is set.

Each call is represented by one transaction; the unit of time chosen is 1 second. It will be assumed that the distribution of arrivals is Poisson with a mean inter-arrival time of 12 seconds. The length of the calls will also be assumed to have an exponential distribution. As in the previous examination of this system, it will be assumed that each new call can come from any of the non-busy lines with equal probability, and that its destination is equally likely to be any line other than itself.

A GPSS block diagram is shown in Fig. 10-5 and the coding in Fig. 10-6. A GENERATE block is used to create a series of transactions representing calls. The modifier at the block is the same function, number 1, used in the supermarket example. The mean of the GENERATE block is set to 12. Parameters 1 and 2 will be used to carry the origin and destination of the call. Each transaction is sent to two ASSIGN blocks to select and record the values. The source of the information is a VARIABLE statement, number 1, which will select a line at random by the

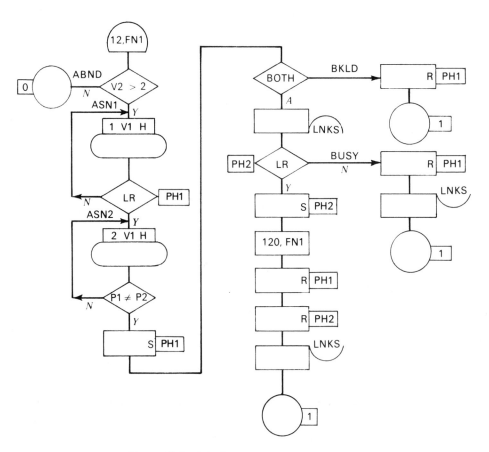

Figure 10-5. Telephone system in GPSS - model 1.

```
* SIMULATION OF TELEPHONE SYSTEM - 1
*
     1       FUNCTION    RN1,C24          FUNCTION FOR EXPONENTIAL DISTRIBUTION
0.0,0.0/0.1,0.104/0.2,0.222/0.3,0.355/0.4,0.509/0.5,0.69
0.6,0.915/0.7,1.2/0.75,1.38/0.8,1.6/0.84,1.83/0.88,2.12
0.9,2.3/0.92,2.52/0.94,2.81/0.95,2.99/0.96,3.2/0.97,3.5
0.98,3.9/0.99,4.6/0.995,5.3/0.998,6.2/0.999,7/0.9997,8
*
 1              GENERATE    12,FN1           CREATE CALLS
 2              TEST G      V2,2,ABND        TEST IF SYSTEM IS FULL
 3      ASN1    ASSIGN      1,V1,PH          PICK ORIGIN
 4              GATE LR     PH1,ASN1         TEST FOR BUSY
 5      ASN2    ASSIGN      2,V1,PH          PICK DESTINATION
 6              TEST NE     PH1,PH2,ASN2     RETRY IF DEST = ORIGIN
 7              LOGIC S     PH1              MAKE ORIGIN BUSY
 8              TRANSFER    BOTH,,BLKD       TRY FOR LINK
 9      GETL    ENTER       LNKS             GET LINK
10              GATE LR     PH2,BUSY         TEST FOR BUSY
11              LOGIC S     PH2              MAKE DEST. BUSY
12              ADVANCE     120,FN1          TALK
13              LOGIC R     PH1              ORIGIN HANGS UP
14              LOGIC R     PH2              DESTINATION HANGS UP
15              LEAVE       LNKS             FREE LINK
16      TERM    TERMINATE   1
        *
17      ABND    TERMINATE   1                ABANDON CALL
        *
18      BLKD    LOGIC R     PH1              ORIGIN HANGS UP
19              TERMINATE   1                BLOCKED CALLS
        *
20      BUSY    LEAVE       LNKS             FREE LINK
21              LOGIC R     PH1              ORIGIN HANGS UP
22              TERMINATE   1                BUSY CALLS
        *
        LNKS    STORAGE     10               NO. OF LINKS
        *
        1       VARIABLE    XH1*RN1/1000+1   PICK A LINE
        2       VARIABLE    XH1-2*S$LNKS-2   COUNT NO. OF FREE LINES
        *
                INITIAL     XH1,50           SET NO. OF LINES
        *
                START       10,NP            INITIALIZE
                RESET
                START       1000             MAIN RUN
```

Figure 10-6. Coding of telephone system - model 1.

method described in Sec. 6-9. The number of lines N is multiplied by a random number between 0 and 1, and the integral part plus 1 is taken to represent a choice of 1 out of N. (The value zero is not allowed for any GPSS entity.) It is not necessary to take any action to extract the integral part because (with certain exceptions) GPSS works with integral numbers. Any evaluation of a VARIABLE statement or function that results in a fractional number is converted to an integer by dropping the fractional part. The number of lines will be varied on different runs so the desired number of lines is placed in halfword savevalue number 1. An INITIAL statement defines the desired value at the beginning of the program. It can appear anywhere in the coding.

The following coding will place line numbers chosen at random from 50 lines in parameters numbers 1 and 2:

	ASSIGN	1,V1,PH
	ASSIGN	2,V1,PH
1	VARIABLE	XH1★RN1/1000+1
	INITIAL	XH1,50

The same variable statement may be used for both assignments because each reference to the VARIABLE will produce a different random number. Note that RN1 is not being used as the input to a function. It therefore is a number in the range 0 to 999 and must be reduced to the range 0 to 1 by being divided by 1,000.

With this method of generating the call origin and destination, it is possible that the origin of the call is already busy. A GATE block checks whether the selected origin is busy by using indirect addressing. If it is, the call is returned for reassignment. Should all lines be busy, this could cause an endless loop, so before assigning the origin, a check is made at a TEST block to ensure that at least two lines are not busy. The TEST block uses VARIABLE 2 to make the check and, if it finds the conditions unsatisfactory, the call is abandoned. It is also possible that the second ASSIGN block will choose the destination to be the same as the origin. This is checked at another TEST block which compares parameter 2 with parameter 1. If they are equal, the transaction is returned to the second ASSIGN block to reassign the destination.

When a valid call has been generated, the model makes the calling line busy by setting a switch, using the SNA PH1. It then checks whether a link is available by attempting to enter the storage called LNKS whose capacity equals the number of links; in this case, 10. If the transaction cannot enter, the call is sent to a TERMINATE block called BLKD and the call is lost after the calling line switch is reset. Transactions that get a link check whether the called party is busy, by using a GATE block. If the line is busy, the call is lost and it goes to a location BUSY where the transaction terminates after resetting the calling line switch and returning the link. Otherwise, the call is established by setting the logic switch corresponding to the destination. An ADVANCE block represents the expenditure of time during the call, using function number 1 with a mean of 120. When the transaction leaves the ADVANCE block, the call is finished and the transaction proceeds to disconnect the call by resetting both logic switches, releasing the link, and terminating.

10-9

Set Operations

An important requirement in a simulation language is the capability of handling sets of temporary entities which have some common property. In GPSS, transactions that are blocked are automatically entered and removed from sets with a FIFO discipline. The program also has a way of controlling sets so that more complex queuing disciplines can be simulated. A number of *user chains* are made available and a transaction is placed on a chain when it enters a LINK block. Field A carries the number (or name) of the chain and field B indicates the queuing discipline. The words FIFO or LIFO result in the disciplines they name. If Pxn is used, the transactions are ordered by ascending values of parameter number n, with a FIFO rule for transactions having the same value. While on the chain, the transactions remain at the LINK block.

To correspond to the LINK block, there is an UNLINK block which allows a transaction (not on the chain) to remove transactions from the chain. The block names the chain in field A, and in field B gives the location to which unlinked transactions are to go. Field C says how many transactions are to be removed. The count can be an integer, the value of an SNA, or the word ALL can be used to remove all the transactions. If only these first three fields are specified, the program removes transactions from the beginning of the chain. However, removal can be made to depend upon the value of any parameter of the transactions on the chain. Field D carries the number of the parameter to be examined and field E carries the value the parameter must have for the transaction to be removed. If field F is used, it provides a location to which the transaction entering the UNLINK block will go if it does not find a transaction on the chain that meets the conditions for removal. If field F is not used, the unlinking transaction goes to the next block, as it always does if it removes a transaction.

As an example of how these blocks are used, suppose that, in the telephone system, blocked calls wait for a link to become free with the following service rules. Line 1 belongs to the company president. If there is an incoming call for line 1 and line 1 is free, the next free link goes to that call. Otherwise, the link goes to the call with the lowest origin number.

The generation of transactions representing calls is the same as before so that the line of blocks on the left-hand side of Fig. 10-5 remains unchanged. The rest of the diagram is replaced by the block diagram of Fig. 10-7. Now, when the calls are blocked because they cannot enter the storage LNKS, they are sent to a LINK block at BLKD. This puts them on a chain called WAIT in ascending order of the origin line number, because the LINK block has 1PH[2] in field B. If the call is successful in getting a link, it checks whether the called line is available, and, if so, the call proceeds as before.

Now, when a call finishes, the program must check whether a call is waiting for the origin or destination lines that have just become free. It begins this checking at the block called CKCH. The check must also be made by any transaction that gets a link but then finds the called line busy. These calls are still going to be lost. Since they release the origin line as they are lost, it is necessary to check the waiting calls. (Note that this call might have been waiting for a link, so it could have been holding the origin line for some time.)

---

[2]See footnote 1, this chapter, regarding the notation used to denote the parameter number.

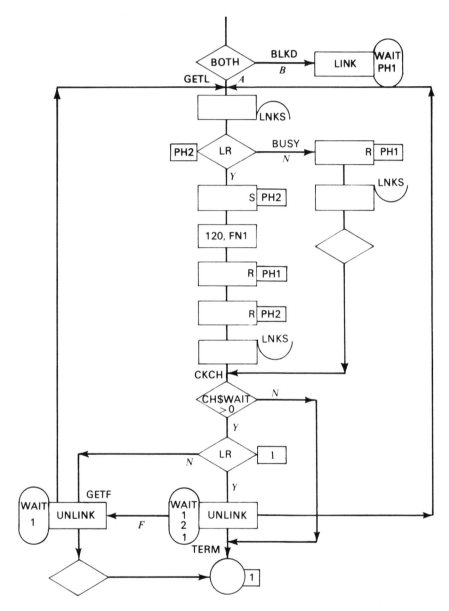

Figure 10-7. Telephone system in GPSS - model 2.

The checking begins with a TEST block examining the SNA called CH$WAIT, which is the number of transactions on the chain WAIT. If there are no waiting calls, no further action is required, and the transaction is terminated. If there is a waiting call, the transaction goes to see if line number 1, the president's line, is free by checking whether logic switch number 1 is reset. If it is, the program is to connect any incoming call to the president. This means searching the chain WAIT for a transaction with parameter 2 equal to 1. The transaction goes to an UNLINK block coded as follows:

UNLINK   WAIT,GETL,1,2PH,1,GETF

This tells the program to search chain WAIT and send to location GETL one transaction such that its second halfword parameter has the value 1.[3] If the transaction fails to find such a transaction, field F instructs it to go to the location GETF. If it is successful in its search, the transaction that was in the UNLINK block goes to the next location, which is the TERMINATE block. A transaction that is unlinked takes over the link that just became free.

If the president's line is busy, or if it is free but there is no call waiting to be connected to it, service is to be offered to the waiting call with the lowest origin number. It will be recalled that when the blocked transactions were put on the chain, they were placed in ascending order of parameter number 1, which is the origin number. The first transaction on the chain, therefore, comes from the lowest origin of the waiting calls. The need is, therefore, to unlink the first transaction of the chain. This is done by the transaction that tried to connect a call to the president's line but failed. It goes to the UNLINK block at GETF, which is coded as follows:

UNLINK   WAIT,GETL,1

Figure 10-8 shows the coding for the problem in its new form. Note that variable 2 has also been changed so that the TEST block, which checks whether it is feasible to generate a new call, now takes account of the waiting calls.

Suppose the system is modified by letting the president have two telephones, on lines 1 and 2, and a call to either is to get first priority. The UNLINK block could be used with another type of SNA called a Boolean variable. These variables are similar to the variables used before but, instead of combining simple SNA's, they use the conditional test phrases of the GATE and TEST blocks. They are distinguished by the operation name BVARIABLE. Each term of a Boolean variable is a single test that could be made by a GATE or TEST block. The terms are combined with the operators ★ and + for AND and inclusive OR, respectively. The individual terms take the value 1 or 0 according to whether the test is true or false. The values are combined by the rules of the operators. For example, consider the following coding:

1   BVARIABLE   PH2'E'1★LR1+PH2'E'2★LR2

---

[3]See footnote 1, this chapter, regarding the notation used to denote the parameter number.

```
*  SIMULATION OF TELEPHONE SYSTEM - 2
*
1        FUNCTION    RN1,C24          FUNCTION FOR EXPONENTIAL DISTRIBUTION
0.0,0.0/0.1,0.104/0.2,0.222/0.3,0.355/0.4,0.509/0.5,0.69
0.6,0.915/0.7,1.2/0.75,1.38/0.8,1.6/0.84,1.83/0.88,2.12
0.9,2.3/0.92,2.52/0.94,2.81/0.95,2.99/0.96,3.2/0.97,3.5
0.98,3.9/0.99,4.6/0.995,5.3/0.998,6.2/0.999,7/0.9997,8
*
```

```
1              GENERATE    12,FN1              CREATE CALLS
2              TEST G      V2,2,ABND           TEST IF SYSTEM IS FULL
3     ASN1     ASSIGN      1,V1,     PH        PICK ORIGIN
4              GATE LR     PH1,ASN1            TEST FOR BUSY
5     ASN2     ASSIGN      2,V1,PH             PICK DESTINATION
6              TEST NE     PH1,PH2,ASN2        RETRY IF DEST = ORIGIN
7              LOGIC S     PH1                 MAKE ORIGIN BUSY
8              TRANSFER    BOTH,,BLKD          TRY FOR LINK
9     GETL     ENTER       LNKS                GET LINK
10             GATE LR     PH2,BUSY            TEST FOR BUSY
11             LOGIC S     PH2                 MAKE DEST. BUSY
12             ADVANCE     120,FN1             TALK
13             LOGIC R     PH1                 ORIGIN HANGS UP
14             LOGIC R     PH2                 DESTINATION HANGS UP
15             LEAVE       LNKS                FREE LINK
16    CKCH     TEST G      CH$WAIT,0,TERM TEST IF CALLS ARE WAITING
17             GATE LR     1,GETF              SEE IF LINE 1 IS FREE
18             UNLINK      WAIT,GETL,1,2PH,1,GETF  CONNECT CALL TO 1
19    TERM     TERMINATE   1
20    GETF     UNLINK      WAIT,GETL,1         CONNECT FIRST WAITING CALL
21             TRANSFER    ,TERM
      *
22    ABND     TERMINATE   1                   ABANDON CALL
      *
23    BLKD     LINK        WAIT,1PH            LINK IN ORDER OF CALL ORIGIN
      *
24    BUSY     LEAVE       LNKS                FREE LINK
25             LOGIC R     PH1                 ORIGIN HANGS UP
26             TRANSFER    ,CKCH               GO TO TEST FOR WAITING CALLS
      *
      LNKS     STORAGE     10                  NO. OF LINKS
      *
      1        VARIABLE    XH1*RN1/1000+1  PICK A LINE
      2        VARIABLE    XH1-2*S$LNKS-CH$WAIT-2   COUNT NO. OF FREE LINES
      *
      *        INITIAL     XH1,50              SET NO. OF LINES
      *
      START    10,NP               INITIALIZE
      RESET
      START    1000                MAIN RUN
```

Figure 10-8. Coding of model 2.

The Boolean variable will be equal to 1 (or true) if parameter number 2 equals 1 and logic switch number 1 is reset, or if parameter number 2 equals 2 and logic switch number 2 is reset. Note that the TEST block conditions are placed between single quotes.

If the first UNLINK block of Fig. 10-7 has BV1 in field D (and nothing in field E), it will unlink transactions that meet the stated condition. With this form of the UNLINK block, it is not necessary to include the GATE block checking for line number 1 to be free.

Two other ways of organizing sets in GPSS will be briefly described, but the block types employed will not be described in detail. Transactions that are on a chain remain static until they are unlinked. It is sometimes necessary to identify members of a set that continue to move around the system. For example, it may be necessary to identify all the jobs in a factory for one customer, or all the cars of a given make. The GPSS program defines a number of *groups* for forming such sets. A block type JOIN allows a transaction to make itself a member of a group and then to proceed to the next block. Another block type REMOVE allows a transaction to remove itself from the group. It can also allow one transaction to remove others from the group, in much the same manner as transactions are unlinked from a chain. It is possible to SCAN the group for particular members, ALTER the parameters of the group members, or EXAMINE a transaction's group membership.

A common reason for wanting to form mobile transaction sets is that they represent inter-related tasks that must be coordinated. The making of a product, for example, may involve many operations, some of which can proceed independently; but some, such as an assembly, require that other operations be finished first. A block type called SPLIT allows one transaction to produce many copies which are automatically linked as a set but may proceed independently. A block type called ASSEMBLE will gather a given number of copies and merge them into one. It is also possible to synchronize the movement of copies with the use of a MATCH block, normally used in pairs. A transaction arriving at a MATCH block must wait until a copy has arrived at another MATCH block. At that time, both transactions can proceed.

Exercises

Draw GPSS block diagrams and write programs for the following problems:

10-1   Customers arrive at a single server counter with an average inter-arrival time of 20 ± 10 seconds. They purchase from 1 to 4 items with the following probabilities:

1	0.5
2	0.2
3	0.2
4	0.1

It takes 5 seconds to purchase each item. Tabulate the distribution of time for serving the first 100 customers.

10-2 Modify the supermarket problem of Fig. 10-2, so that 10% of the customers do not use a basket but otherwise are the same as other customers.

10-3 Jobs are passed into an office at the rate of one every 15 ± 5 minutes. Normally they go to clerk A, who takes 10 ± 3 minutes and then go to clerk B, who takes 5 ± 2 minutes. However, if clerk B is busy at the time the job is brought to the office, the entire job is given to clerk C, who takes 20 ± 10 minutes. Find out how many of the first 1,000 completed jobs will have been handled by clerk C. Assume each clerk handles one job at a time and assume that work can be stacked between clerks A and B.

10-4 People arrive at a bus stop at the rate of one every 20 ± 15 seconds. They queue for the bus unless the queue already has 10 people, in which case they walk away. A bus arrives every 5 ± 1 minutes. The people waiting board the bus, one at a time, taking 5 seconds each. The bus waits for at least 20 seconds, otherwise it leaves as soon as the people stop boarding. Simulate 10 busloads of people. (Use separate transaction streams to represent people and buses and give a higher priority to the transactions representing the people. Use a logic switch to represent the presence or absence of a bus.)

10-5 Messages are being generated at the rate of one every 7 ± 3 seconds. They are to be transmitted to 1 of 4 destinations with the following probabilities:

1	0.2
2	0.3
3	0.35
4	0.15

All messages are first sent by a single main line to a switching center which then sends them on to their destinations by way of individual lines. Only one message can be on each of the lines at any time. If necessary, the switching center can store messages to await a free line. The messages are uniformly distributed in size from 10 to 100 characters. The main line can send messages at the rate of 10 characters a second while the other lines can only send them at the rate of 5 characters a second. Simulate the transmission of 1,000 messages and measure the queues on each line. (Use parameters to carry the destination.)

10-6 In the telephone system model of Fig. 10-5, assume that 20% of the calls are long distance calls; the rest are local calls. Long distance calls are charged $1 each and local calls are charged 10¢ each. No charge is made for calls not connected. Modify the program to measure the telephone company's revenue.

10-7 A segment of a railroad consists of a single track section followed by a double track section. Only one train at a time can enter the single track section. As each train clears the single track section, a switch is thrown so that alternate trains go to each section of the double section. There is no limit on how many trains can be in the double train track section. Assume trains arrive every 10 ± 5 minutes and take 8 ± 4 minutes to clear the single track and 18 ± 9 to clear the double track section. Simulate the movement of 1,000 trains through the system and measure the queue that forms for the single track.

11

INTRODUCTION TO SIMSCRIPT

11-1

SIMSCRIPT Programs

SIMSCRIPT is a very widely used language for simulating discrete systems. Beginning with two early versions, (1) and (5), now obselete, it has progressed through two major, intermediate versions, SIMSCRIPT I.5 (2), and SIMSCRIPT II, (3) and (6), to the current and most powerful version, SIMSCRIPT II.5, (4). It has been implemented on several different manufacturers' computers. SIMSCRIPT II is generally compatible with SIMSCRIPT II.5. However, there are some important differences, some of which are dependent on the particular machine implemention. The differences are described in the preface of Ref. (4), which is the second edition of Ref. (3). The description given here will be of the II.5 version.[1] For brevity, however, the program will be called, simply, SIMSCRIPT.

The language can, in fact, be considered as more than just a simulation language since it can be applied to general programming problems. The description of the language given in Refs. (3) and (4) is organized in five levels. Beginning with a simple teaching level to introduce the concepts of programming, the description

---

[1] A modification of SIMSCRIPT II.5 has been announced that allows the program to be used in a new manner, as an alternative. The description given here and in the next chapter will be for the program operating in its traditional manner, as described in Ref. (4). The general nature of the alternative mode of operation will be explained in Sec. 13-8.

adds levels corresponding to a scientific programming language, comparable to FORTRAN; a general purpose language, comparable to ALGOL or PL/I; a list processing language, needed for creating the data structures of a simulation model; and, finally, the simulation-oriented functions needed to control the simulation and gather statistics. The description given here will be of the full language. The present chapter will cover the simpler telephone system model that was described in Chap. 8, and programmed as model 1 in GPSS. This model is for a lost-call system and does not have any sets. The next chapter will discuss model 2 of the telephone system, which includes sets of calls waiting for connection.

11-2

SIMSCRIPT System Concepts

The viewpoint to be taken by the SIMSCRIPT user corresponds essentially to that used in Sec. 1-1 to discuss the general nature of systems. That is to say, the system to be simulated is considered to consist of entities having attributes that interact with activities. The interaction causes events that change the state of the system. In describing the system, SIMSCRIPT uses the terms *entities* and *attributes*. For reasons of programming efficiency, it distinguishes between *temporary* and *permanent* entities and attributes. The former type represents entities that are created and destroyed during the execution of a simulation, while the latter represents those that remain during the run. A special emphasis is placed on the manner in which temporary entities form sets. The user can define *sets*, and facilities are provided for entering and removing entities into and from sets.

Activities are considered as extending over time with their beginning and (usually) their end being marked as events occurring instantaneously. Each type of event is described by an *event routine*, each of which is given a name and programmed as a separate closed subroutine. A distinction is made between endogenous (or internal) events, which result from actions within the system, and exogenous (or external) events, which arise from actions in the system environment. An endogenous event is caused by a scheduling statement in some event routine. The event marking the beginning of an activity will usually schedule the event that marks the end of the activity. If the beginning or ending of an activity implies the beginning of some other activity, then the event routine marking the beginning or end of the first activity will schedule the beginning of the second.

Exogenous events require the reading of data supplied by the user. Among the data is the time at which the event is to occur. There can be many data sets representing different sets of external events. Event routines are needed to execute the changes that result when an external event becomes due for execution. Part of the automatic initialization procedure of SIMSCRIPT is to prepare the first exogenous event from each data set.

In practice, external events, such as arrivals, are often generated using the "bootstrap" method that was described in Sec. 8-3. Given the statistical distribution of the inter-arrival time (together with the statistical properties of any attributes that need to be represented) an endogenous event routine continuously creates one arrival from its predecessor. An exogenous event routine is essential only when specific data are needed, as in the cases where historical data are being used, or several runs are being made with exactly the same data in order to minimize variance between runs comparing different system designs.

In summary, the concepts used in SIMSCRIPT are

> Entities
> > Permanent
> > Temporary
> Sets
> Event routines

11-3
Organization of a SIMSCRIPT Program

Since the event routines are closed routines, some means must be provided for transferring control between them. The transfer is effected by the use of *event notices* which are created when it is determined that an event is scheduled. At all times, an event notice exists for every endogenous event scheduled to occur, either at the current clock time or in the future. Each event notice records the time the event is due to occur and the event routine that is to execute the event. If the event is to involve one of the temporary entities, of which there may be many copies, the event notice will usually identify which one is involved.

The general manner in which the simulation proceeds is illustrated in Fig. 11-1. The event notices are filed in chronological order. When all events that can be executed at a particular time have been processed, the clock is updated to the time of the next event notice and control is passed to the event routine identified by the notice. These actions are automatic and do not need to be programmed. Event notices do not usually go to more than one activity in the manner of a GPSS transaction. Having activated the routine, the event notice has served its purpose and is usually destroyed. However, it is possible to save and reschedule an event notice.

If the event executed by a routine results in another event, either at the current clock time or in the future, the routine must create a new event notice and file it

TEMPORARY ENTITY RECORDS

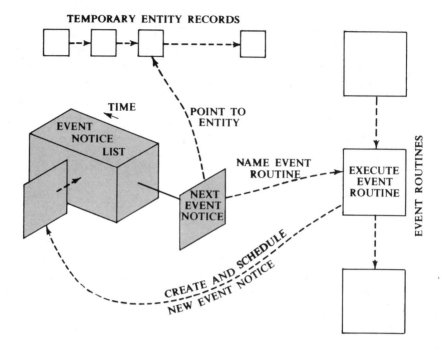

Figure 11-1. SIMSCRIPT execution cycle.

with the other notices. For example, in the telephone system, the connection of a call implies its disconnection at a later time, so the routine responsible for connecting the call will be responsible for producing the event notice that schedules the disconnection.

In the case of the exogenous events, a series of *exogenous event statements* are created; one for each event. The statements are similar to event notices in that they give the time the event is to occur and identify the exogenous routine to execute the event. All exogenous event statements are sorted into chronological order and they are read by the program when the time for the event is due.

11-4

Names and Labels

The user must describe all the entities by giving a name to each entity and its attributes. Names may consist of any combination of letters and digits provided there is at least one letter. (Some names are reserved for system use.) It is also permissible to use periods in a name so that compound names can be constructed. For example: if an entity type is to represent persons, it could be named PERSON;

if age and education are to be attributes, they could be named AGE and EDUCA-TION, or they could be named PERSON.AGE and PERSON.EDUCATION. In practice, the specific machine implementation of the language will only recognize a certain number of initial characters (typically eight) within any name, but for documentation purposes the name can be of any length.

Labels for identifying programming statements similarly consist of any combination of letters and numbers, without the restriction that at least one be a letter. They also may be made compound by using periods. Labels are identified by being enclosed between single quotation marks.

11-5

SIMSCRIPT Statements

SIMSCRIPT statements are written in a form closely resembling the English language. To enhance the similarity, there are many alternative terms or modes of expression which will be interpreted as equivalent statements. For example, a statement calling for n lines of text to be printed could be written in either of the following ways:

PRINT n LINES AS FOLLOWS
PRINT n LINES THUS

The n lines of text then follow, exactly as they are to be printed. If the values of some variables are to be included in the text, the statement could read as follows:

PRINT n LINES WITH X AND Y LIKE THIS

The symbols X and Y represent the variables or expressions to be printed. The lines of text that follow the statement will indicate with asterisks and, if necessary, decimal points where the numbers are to be printed. A three digit integer, for example, would be indicated by ★★★, and a four digit real number with one decimal place would be indicated by ★★★.★. In this short account of SIMSCRIPT no attempt will be made to point out all the alternative forms that can be taken by statements.

Reading the program can be made easier by including comments, indicated by beginning a statement with two single quotes (not a double quote mark). The comment can be concluded by another pair of single quotes, but, if the entire line is to be nothing but a comment, the trailing quotes are not necessary. Comments can, in fact, be included within a statement, in which case it is necessary to mark the end of the comment. For example, a printing statement might say

PRINT "AS THE REPORT HEADING" n LINES THUS

Blank spaces may also be freely used to make program listings easier to read, by indenting statements or spacing words.

Table 11-1 lists the SIMSCRIPT statements that will be used in programming the event routines. Some of them will not be described until the next chapter.

TABLE 11-1 SIMSCRIPT Statements

CREATE   *temporary entity* or *event notice* (CALLED *variable*)
DESTROY

SCHEDULE AN *event notice* (CALLED *variable*) $^{IN}_{AT}$ *expression*

LET *variable* = *expression*

FILE *variable* IN *set*

REMOVE $^{FIRST}_{LAST}$ *variable* FROM *set*

FOR *variable* = *expression.1* TO *expression.2* BY *expression.3*

FOR EACH *permanent entity*

FOR EACH *variable* OF *set*

WITH *expression.1 comparison expression.2*
OR
AND
UNTIL

IF *expression.1 comparison expression.2, statement*

IF *set* $^{IS}_{IS\ NOT}$ EMPTY, *statement*

GO TO *statement label*

DO

LOOP

FIND THE FIRST CASE, IF NONE *statement* ELSE

11-6

Defining the System

A SIMSCRIPT program begins with a preamble section that defines the system structure and establishes the conditions under which the simulation will be run. Variables can be real, integer, or alphabetical. The size depends upon the particular machine implementation and on whether single or double precision is chosen. It is also possible to pack variables so that more than one value is placed in one word. In addition, it is possible to process character strings, in what is called the TEXT mode of operation. If no specific declaration is made, the program operates with single precision in a real mode. The statement

NORMALLY, MODE IS INTEGER (or ALPHA)

changes the mode. Whichever mode is established as normal, individual variables can be defined as being in another mode.

Names must be given to every entity and its attributes. Also, the simulation will need to define certain words or tables for such purposes as carrying out computations, gathering statistics, or holding initialization values. If these are to be global variables (that is, available to all routines) they are called *system variables*. Those that are to be in the normal mode of the program can be listed after the statement

THE SYSTEM HAS

Items in this or any other SIMSCRIPT list can be separated by commas or the word AND (or both). If the mode is to be other than the normal mode, a DEFINE statement lists the variables and declares their mode, as follows:

DEFINE *name.1,name.2* . . . AS INTEGER VARIABLES

If the DEFINE statement does not mention the mode, the variables are taken to be in the normal mode.

Events are represented by individual routines. Within each of these event routines it is possible to use local variables, accessible only to that routine. These must not be defined in the preamble. They do not, in fact, have to be specifically defined in the event routine, unless their mode differs from the normal mode, in which case the DEFINE statement is used within the event routine.

The permanent entities, temporary entities, and the event notices are defined in lists, following the statements PERMANENT ENTITIES, TEMPORARY ENTITIES, and EVENT NOTICES, respectively. Following the opening statement is a line for each entity, taking the following form:

EVERY *entity* HAS A *attr.1* AND A *attr.2* . . .

where *entity* is the name of the entity and *attr.i* is the name of its *i*th attribute. There can be any number of attributes. If any SIMSCRIPT statement needs to go beyond 80 characters it can be extended to another line.

Event notices have the same definition format as temporary entities, although a number of words within the record are automatically reserved for use by the program. The event notices can also carry data, supplied by the user, just like the attributes of temporary entities. If they are needed, the attributes are defined with a list of EVERY statements following an EVENT NOTICES statement, just as for permanent or temporary entities. Often there is no need for data other than the automatically defined system data. In that case, definition of the event notices can be simplified by listing the names of those not needing user-defined attributes, in a sequence, after the statement

EVENT NOTICES INCLUDE...

If an array or multidimensional table is needed, a **DEFINE** statement, giving the name of the entity, defines the dimensions, in addition to declaring the format, if that is necessary. For example:

DEFINE *table* AS AN INTEGER, *n*-DIMENSIONAL VARIABLE

where *table* is the table name, and *n* is the number of dimensions (for which there is no specific limit.)

There are other controls that can be exercised by statements in the preamble of a SIMSCRIPT program, such as whether single or double precision arithmetic is to be used. One control that is often needed is to redefine the name of the time unit. If no specification is made, units of days, hours, and minutes are used by following any specification of a time with one of the words **DAYS, HOURS,** or **MINUTES.**

11-7

Defining the Telephone System Model

To illustrate the use of SIMSCRIPT, the telephone system that was described in Chap. 8, and later programmed in GPSS, will be programmed as a SIMSCRIPT model. (See Sec. 8-4 for a description of the system.) The program will be for a lost-call system, in which calls that are unable to make a connection immediately are abandoned. The next chapter will expand the model.

The telephone lines are obviously permanent entities, and each needs an attribute indicating whether the line is busy or not. The name **TLINE** will be used to represent telephone lines, and the attribute will be called **STATE.** It is an integer variable, and, in fact, takes only the values 0 or 1, to represent a free or busy state. Calls are to be represented by temporary entities, each with two attributes for carrying the origin and destination.

As in the previous implementations of this model, the links will not be represented individually. It is not necessary, therefore, to define a permanent entity for their representation. Instead, two system variables will be needed to carry the maximum number of links and the number currently in use. The maximum number will be initialized from data read into the program. So also will be the mean inter-arrival time, the mean call length, and the time the simulation is to run, and so system variables need to be defined for these quantities. Other system variables will be used to collect statistics on how many calls are processed, and how many are abandoned because they are blocked or find the called party busy.

The preamble for model 1 of the telephone system is shown in Fig. 11-2. The figure actually shows part of the printed output of a compilation, which lists the input. The line numbers shown at the left in the figure are placed there by the compilation: they are not entered by the user.

```
 1   ''   TELEPHONE SYSTEM - MODEL 1
 2   PREAMBLE
 3   NORMALLY,MODE IS INTEGER
 4   EVENT NOTICES INCLUDE ARRIVAL AND CLOSING
 5        EVERY DISCONNECT HAS A DIS.CALL
 6   TEMPORARY ENTITIES
 7        EVERY CALL HAS AN ORIGIN AND A DESTINATION
 8   DEFINE LINKS.IN.USE, MAX.LINKS, BLOCKED, BUSY, FINISHED
 9        AND STOP.TIME AS VARIABLES
10   DEFINE INTER.ARRIVAL.TIME AND MEAN.LENGTH AS REAL VARIABLES
11   PERMANENT ENTITIES
12        EVERY TLINE HAS A STATE
13        DEFINE SECS TO MEAN /60 MINUTES
14   END
```

Figure 11-2.   Telephone system - model 1, preamble.

Line 1 is simply a comment. Because it contains nothing but a comment, it is not essential to use the two quote marks at the end of the comment. Only the inter-arrival time and the mean call length need to be specified as real variables, so it is convenient to define the normal mode as being integer. That is done by the statement of line 3.

As will be explained shortly, three event routines will be needed, therefore there must also be three event notices. Lines 4 and 5 define the event notices. By convention the names of the event notices must be the same as the event routines they initiate. In this case, the routines are going to be called ARRIVAL, DISCONNECT, and CLOSING. Because the ARRIVAL and CLOSING event notices do not need attributes, their definition is included in the event notice header. The DISCONNECT routine, however, will need to know which particular call is to be disconnected, so the DISCONNECT event notice needs an attribute with which to identify the call.

Lines 6 and 7 define the temporary entity representing the calls. It has two attributes for recording the origin and destination of the call. The next three lines of the listing define the system variables, some as integers (by implication) and some as real variables. Lines 8 and 9 illustrate how statements can be carried over more than one line. Lines 11 and 12 define the permanent entities representing the telephone lines. Line 13 defines a time unit of seconds, and line 14 is an obligatory END statement for indicating the end of the preamble section.

The purposes of the system variables are apparent from their names. In addition, there are many system variables automatically provided by the program.

These carry names defined by the system and do not have to be specified in the preamble. In particular, every permanent entity definition implies a system variable called N.*name*, where *name* is the name given to the permanent entity by the user. The variable holds the number of permanent entities of that type, and a value must be read into the program at the time of intialization.

11-8
Referencing Variables

Single variables, such as the unique system variables, are, of course, referenced by using their names. Permanent entities, such as the telephone lines, are represented by arrays, with an array for each attribute, identified by the attribute name. Individual entities are identified by an integer number which is an index to the array. An array defined by the user as a system variable is similarly entered by using an integer index. Reference to the state of a particular telephone line, for example, is made with the term STATE(I), where *I* is an integer or an expression that can be evaluated as an integer. If an expression is used, it may also be indexed, and such indexing can continue to any depth. For example, in the present model there will be a need to use the term STATE(ORIGIN(CALL)). This refers to the state of a line identified by the origin attribute of the event notice called CALL.

Temporary entities and event notices are created and destroyed as the simulation proceeds. The simplest method names only the type of entity or event notice involved. For example, CREATE A CALL, or DESTROY A CALL. For each creation, the program refers to the data structure given in the preamble, and creates a block of words. Since the number of blocks will fluctuate, they cannot be kept in a fixed table. Instead, a pointer is used to give the location of the block. Whenever the term "name" is used in reference to a temporary entity or event notice, it means the location of the pointer identifying the entity or notice.

For every type of temporary entity, or event notice, the system automatically reserves a location for holding a pointer that has the same name as the entity or notice. If the simple (or short) form of the CREATE statement is used, the pointer to the entity or notice being created is placed in that location. The command CREATE CALL, for example, puts the pointer to the created record in a location called CALL.

If, as will happen in the present model, it is necessary to identify that particular entity at some later time, the content of that location must be saved for future reference. In the present example, the event notice that will schedule the disconnection of the call will carry the content of CALL as an attribute, so that the routine that is to perform the disconnection can later identify the call.

If, during the current execution of a routine, it is necessary to create a second copy of an entity or notice, the second creation will write over the pointer to the first unless another (long) form of the **CREATE** statement is used. This takes the form **CREATE A CALL CALLED** *name*. The pointer to the created record then goes to the location called *name*. Different names can be used for different creations, or, since *name* can be a variable, the creations can be under index control. The **DESTROY** statement and the **SCHEDULE** statement, to be described later, can also take a long form.

11-9

The MAIN Routine

Every SIMSCRIPT program has a section called MAIN, which controls the sequence of tasks needed to initialize the simulation, carry out the execution, and prepare outputs. The MAIN routine for the present telephone system example is shown in Fig. 11-3. It immediately follows the preamble.

Line 1 identifies the program section with the word MAIN. The first task needed, in the present case, is to initialize a run. The statement at line 2 reads in five values. A record containing the values is prepared in free format, which allows the values—in the format required by the definition of the preamble—to be entered with one or more blanks separating them. The example of Fig. 11-3 is for a single run only. If

```
 1   MAIN
 2   READ N.TLINE, MAX.LINKS, INTER.ARRIVAL.TIME, MEAN.LENGTH AND STOP.TIME
 3   CREATE EVERY TLINE
 4   IF N.TLINE < 2*MAX.LINKS + 2
 5        PRINT 1 LINE THUS
    TOO FEW LINES SPECIFIED. SIMULATION ABANDONED.
 6        STOP
 7   ELSE
 8   SCHEDULE AN ARRIVAL NOW
 9   SCHEDULE A CLOSING IN STOP.TIME MINUTES
10   START SIMULATION
11
12   ''   REPORT TO BE WRITTEN AT END OF SIMULATION
13
14        PRINT 1 LINE THUS     '
SIMULATION OF TELEPHONE SYSTEM - MODEL 1
15        SKIP 3 OUTPUT LINES
16        PRINT 5 LINES WITH N.TLINE, MAX.LINKS, INTER.ARRIVAL.TIME,
17             MEAN.LENGTH, STOP.TIME THUS
    NUMBER OF LINES              *
    NUMBER OF LINKS              *
    MEAN INTER-ARRIVAL TIME      *.*  /SECONDS
    MEAN CALL LENGTH             *.*  /SECONDS
    SIMULATED TIME               *    /MINUTES
18        SKIP 2 OUTPUT LINES
19        PRINT 3 LINES WITH FINISHED, BUSY, BLOCKED THUS
    CALLS PROCESSED              *
    BUSY CALLS                   *
    BLOCKED CALLS                *
20        STOP
21   END
```

Figure 11-3. Telephone system - model 1, MAIN routine.

the simulation were to be run for many cases, programming could be established to read an input and execute a simulation repeatedly. Among the inputs are numbers establishing the sizes of the arrays representing the permanent entities, in this case, the number N.TLINE for determining the number of telephone lines. The CREATE statement at line 3 is needed to establish the array. (In other circumstances, the value might have been calculated within the routine MAIN, rather than just being read as input.)

The method to be used for selecting lines for the origin and destination of a call is to pick a line number at random, with a uniform distribution, assuming that the choice is legal. As explained in Sec. 10-8, which gives the GPSS implementation of this model, the number of lines must exceed twice the number of links, or an endless loop is possible. Lines 4 through 7 illustrate how conditions can be programmed by making the test relating the numbers of lines and links. The statement of line 4 is an IF statement which is evaluated to be either true or false. If it is true, the program executes the statement directly following. In this case, if the number of lines is less than twice the number of links plus two, the program executes line 5. This calls for the printing of a remark, and leads to line 6 which stops the simulation. If the IF statement is found to be false, the program advances to the next ELSE statement, and continues execution from that point.

In the present case, this leads to executing line 8 which is a SCHEDULE statement, calling for the creation of an ARRIVAL event notice. A SCHEDULE statement must specify the time at which the event is to occur. Line 8 specifies an immediate execution. However, program control at this point still resides in the MAIN routine. The SCHEDULE statement will file the event notice for later execution. The next statement, at line 9, is also a SCHEDULE statement, this time calling for a CLOSING event at the STOP.TIME that was read as part of the input. This event notice will also be filed and, since event notices are kept in chronological order, it will be placed after the ARRIVAL event notice.

The simulation has now been initialized, and the MAIN routine calls for execution of the simulation by going to the statement START SIMULATION. Control will then pass from the MAIN routine to an internal routine that executes the events of the simulation. For this model, only two events have been scheduled at this point, but executing the first arrival will cause other events to be created. The simulation will proceed by executing events until the time for the CLOSING event is reached, to end the simulation. At that point control returns to the MAIN routine at the statement following the one that started the simulation.

An output report is to be written at this point. Lines 14 through 19 define the report. An explanation of how these statements are executed was given in Sec. 11-5. In fact, it is evident from reading the statements how the report will appear since they simply indicate lines as they are to be printed, with the names of the variables that are to be included. Figure 11-4 shows the output produced by running this

```
SIMULATION OF TELEPHONE SYSTEM - MODEL 1


    NUMBER OF LINES                    50
    NUMBER OF LINKS                    10
    MEAN INTER-ARRIVAL TIME          12.0  /SECONDS
    MEAN CALL LENGTH                120.0  /SECONDS
    SIMULATED TIME                   200  /MINUTES


    CALLS PROCESSED                   988
    BUSY CALLS                        218
    BLOCKED CALLS                      76
```

Figure 11-4. Telephone system - model 1, output report.

simulation. The output lists the conditions under which the simulation was run. When the report has been written, the simulation is stopped by the statement at line 20. As with all SIMSCRIPT routines, an END statement is needed to indicate the physical end of the routine definition.

11-10

The Arrival Event

The routine for executing an arrival event is shown in Fig. 11-5. As with all event routines, it begins with a statement naming the event routine, and it ends with an END statement. The user writing the program now assumes control has been passed to this routine, and programs the changes that are to occur as a result of the event being executed. During the execution of the routine the clock will stand still.

First, the routine creates a record representing the call with the statement CREATE CALL. The program refers to the information given in the preamble for the record format. Lines 3 through 8 select an origin and destination for the call and place the values in the attributes of the call record. The selection is made at random from among all lines, with a uniform distribution. The right-hand side of statement 3 is a SIMSCRIPT function for generating a uniformly distributed integer—in this case, between 1 and the number of lines, N.TLINE. (The third parameter of the calling sequence selects a random number generator.) The selected origin, however, must be a line that is not currently busy, and the destination must not be the origin. The program repeats the selections if these conditions are not met (using a structured programming technique, rather than explicitly programming a loop.)

The statement of line 9 then checks whether the call can be connected. It is a compound IF statement requiring that there be a free link and that the called party

```
 1  EVENT ARRIVAL
 2       CREATE CALL
 3       LET ORIGIN(CALL) = RANDI.F(1,N,TLINE,1)
 4       UNTIL STATE(ORIGIN(CALL)) = 0,
 5            LET ORIGIN(CALL) = RANDI.F(1,N,TLINE,1)
 6       LET DESTINATION(CALL) = RANDI.F(1,N,TLINE,1)
 7       UNTIL DESTINATION(CALL) NE ORIGIN(CALL),
 8            LET DESTINATION(CALL) = RANDI.F(1,N,TLINE,1)
 9       IF STATE(DESTINATION(CALL)) = 0 AND LINKS.IN.USE < MAX.LINKS,
10            LET STATE(ORIGIN(CALL)) = 1
11            LET STATE(DESTINATION(CALL)) = 1
12            ADD 1 TO LINKS.IN.USE
13            SCHEDULE A DISCONNECT GIVEN CALL
14                 IN EXPONENTIAL.F(MEAN.LENGTH,1) SECS
15       ELSE
16            IF LINKS.IN.USE = MAX.LINKS,
17                 ADD 1 TO BLOCKED
18            ELSE
19                 ADD 1 TO BUSY
20            ALWAYS
21            ADD 1 TO FINISHED
22            DESTROY CALL
23       ALWAYS
24       SCHEDULE AN ARRIVAL IN EXPONENTIAL.F(INTER.ARRIVAL.TIME,1) SECS
25       RETURN
26  END
```

Figure 11-5. Telephone system - model 1, arrival event.

is free for the statement to be true. When it is true, so that a call can be connected, lines 10 through 14 are executed. These statements make changes to the system image corresponding to the connection of a call. The two lines involved in the call are made busy and the number of links in use is increased by one. Lines 13 and 14, which are one statement, schedule an event to represent the eventual disconnection of the call. It is a SCHEDULE command which names the DISCONNECT event notice as the one to be created and filed. The time for the event is to be the present time plus a random number selected from an exponential distribution whose mean was read at the time of initialization. The term EXPONENTIAL.F makes reference to an internal SIMSCRIPT function that creates such a random number. (Again, the extra parameter of the calling sequence specifies a random number generator.) The use of the word IN implies the addition of the generated number to the current clock time. Using the word AT would have made the generated time the actual time. The SCHEDULE statement includes the phrase GIVEN CALL. This places the pointer identifying the particular call in the attribute DIS.CALL of the event notice that is being scheduled. In this way the DISCONNECT routine, when it is executed, will be able to identify which call it is being asked to disconnect.

If the conditional statement at line 9 is not true, so that the call cannot be connected, the program moves to the section following the ELSE statement at line 15. The next four lines of programming determine whether the reason for the call not being connected is that there is no link or that the called party is busy. One or other of the counters BLOCKED or BUSY is incremented accordingly.

The statement ALWAYS is used to mark the end of all the statements covering the alternatives involved in an IF statement, so that, whichever condition occurs, the program will eventually come to the ALWAYS statement associated with the IF statement. In the present example, there is one IF statement at line 16 deciding which condition caused a call to be abandoned. The ALWAYS statement corresponding to *that* IF statement is at line 20. It makes the program execute line 21, whatever the cause of abandoning the call. The counter FINISHED will therefore be incremented when a call is abandoned, so that the number of calls being reported in the output as processed includes these calls, along with calls that *are* connected and completed by the time the simulation ends. After noting that the call is to be abandoned, the program removes this call from the simulation with a DESTROY CALL statement.

There is a second ALWAYS statement, at line 23, which is associated with the IF statement at line 9. The associations between the sets of IF and ALWAYS statements is made by noting how the pairs of statements are nested, one within the other. (Note how indentation of the program listing has been used to help the reader make these associations.) The program will come to the ALWAYS statement at line 23 whether the call is connected or not. The purpose is to execute the statement at line 24.

In the present example, the bootstrap method of generating one arrival from its predecessor is being used. Line 24 schedules the next arrival for a time ahead of the current clock time selected from an exponential distribution. The last executable statement of the event routine is a RETURN statement which relinquishes control to the timing routine.

11-11

The Timing Routine

Following the initial arrival of a call there will be two event notices (in addition to the CLOSING event notice.) One will be for the next arrival and one for the disconnection of the first arrival. They could occur in either order. Later executions of the ARRIVAL routine will similarly produce two more event notices. The event notices are filed in chronological order. Whenever an event routine relinquishes control, the program automatically returns to a section of the program that selects the next event notice. The clock is updated to the time of that event (which might be the current clock time if simultaneous events occur) and control is passed to the event routine associated with the event notice. If there should be no event notices, the simulation algorithm assumes the simulation is complete and relinquishes control to the MAIN routine.

11-12

The Disconnect Event

When the program selects a disconnection as the next event, control is transferred to the routine shown in Fig. 11-6. There are no conditions to be met before a call is disconnected, once its time of disconnection has arrived. With this simple lost-call model of a telephone system there are, also, no calls waiting upon the completion of existing calls, so the routine is relatively simple.

```
1   EVENT DISCONNECT GIVEN CALL
2       LET STATE(ORIGIN(CALL)) = 0
3       LET STATE(DESTINATION(CALL)) = 0
4       DESTROY CALL
5       SUBTRACT 1 FROM LINKS.IN.USE
6       ADD 1 TO FINISHED
7       RETURN
8   END
```

Figure 11-6. Telephone system - model 1, disconnect event.

The two lines involved in the call are made free. To do this the program has to use the attribute of the event notice to identify the call. Reference to the lines is made by the use of double subscripts that was described in Sec. 11-8. The call is then destroyed. (Note the actions must be done in this order so that the program has the reference to the call when it alters the status of the lines.)

The only other action needed is to increment the count of the calls that have been finished. This is an example of an event that does not imply another event, so there is no creation of an event notice within the routine.

11-13

The Closing Event

The simplest (and customary) way of terminating a simulation is to arrange that the program runs out of events to execute. In the present model, the intention is to use the stopping time that was read into the program as the time at which new calls stop arriving. Calls in the system at that time are to be completed.

When the closing event notice, which was created in the MAIN routine, becomes due, control is passed to the routine shown in Fig. 11-7. Because of the way arrivals are being created, there is only one event notice for an arrival at any time. The first executable statement of the routine says CANCEL ARRIVAL, which will remove the arrival event notice from the set of notices scheduled for future execution. The notice is not automatically destroyed as a result of being

```
1  EVENT CLOSING
2      CANCEL THE ARRIVAL
3      DESTROY THE ARRIVAL
4      RETURN
5  END
```

Figure 11-7. Telephone system - model 1, closing event.

cancelled, so that it would be possible to reschedule it at some later time. However, in the present model the notice is destroyed by the statement that follows the cancellation.

The routine then returns control to the timing routine. The cancellation stops further arrival of calls but does not necessarily stop the simulation because there could be calls in progress. If so, the timing routine will continue to select the disconnection of these calls until none is left. The absence of further events will then cause the program to relinquish control to the MAIN routine, leading to the printing of the output report that has already been described.

Exercises

11-1   Explain the changes that must be made to the telephone system model to introduce the following modifications. Identify clearly whether the changes involve entities, attributes, or activities. Treat each change separately.

(a) The program is to stop at clock time 50,000.

(b) Random breakdowns in service periodically cut off all calls in progress. Service is restored immediately.

(c) The accumulated length of all calls is to be recorded.

(d) Records are to be kept of all the individual link occupancies.

11-2   Change the telephone system simulation so that it takes 10 ± 5 time units to connect a call or find out that the called line is busy. A link is held during this period.

11-3   Change the telephone system simulation to charge calls at the rate of 1¢ per time unit and record the total cost of calls.

11-4   Change the telephone system simulation so that 10% of the calls are for weather or time information. These calls do not need a link and they can always be connected. Their length is 20 ± 10 time units.

11-5   Parts are produced by a machine tool at the rate of one every five minutes. Each part is inspected for 4 ± 3 minutes and 10% are rejected. Write a SIMSCRIPT program to simulate the system.

11-6   People arrive at a bus stop with inter-arrival times of 3 ± 1 minutes. A bus arrives with inter-arrival times of 15 ± 5 minutes. The bus has a capacity of

30 people, and the number of seats occupied when the bus arrives is equally likely to be any number from 0 to 30. The bus takes on board as many passengers as it can seat and passengers that cannot be seated walk away. Simulate the arrival of 100 buses and count how many people do not get on board.

Bibliography

1   DIMSDALE, B., AND H. M. MARKOWITZ, "A Description of the SIMSCRIPT Language," *IBM Syst. J.*, III, no. 1 (1964), 57–67.

2   KARR, HERBERT W., HENRY KLEINE, AND HARRY M. MARKOWITZ, *SIM-SCRIPT I.5*, Santa Monica, Calif.: California Analysis Center, Inc., 1966.

3   KIVIAT, P. J., R. VILLANUEVA, AND H. M. MARKOWITZ, *The SIMSCRIPT II Programming Language*, Englewood Cliffs, N. J.: Prentice-Hall Inc., 1969.

4   ———, (Ed. E. C. RUSSELL), *SIMSCRIPT II.5 Programming Language*, Los Angeles, Calif.: CACI, Inc., 1975.

5   MARKOWITZ, HARRY M., BERNARD HAUSNER, AND HERBERT W, KARR, *SIMSCRIPT-A Simulation Programming Language*, Englewood Cliffs, N. J.: Prentice-Hall, Inc., 1963.

6   WYMAN, FORREST PAUL, *Simulation Modeling: A Guide to Using SIMSCRIPT*, New York: John Wiley & Sons, Inc., 1970.

12

MANAGEMENT OF SETS
IN SIMSCRIPT

12-1

Definition of Sets in SIMSCRIPT

SIMSCRIPT allows the formation of sets linking groups of temporary entities having a common property. The definition is made by statements in the preamble of the program. One type of statement names the set and defines the discipline by which it is organized. Another type specifies who owns the set.

Sets can be defined with subscripts. Without a subscript, the set is declared as belonging to the system, using the statement

THE SYSTEM OWNS A set

A single-subscript set definition associates one set with every member of some entity. The association is made in the EVERY statement defining the attributes of the entity type, in the following manner:

EVERY entity OWNS A set AND HAS AN attribute

A double-subscripted set definition associates one set with every pair of two entity types. An ownership statement appears in the preamble, naming the two entity types connected with the same set.

264

For example, in the telephone system, if blocked calls wait for a link to become free, an unsubscripted set will be defined to hold waiting calls. If busy calls wait for a line to become free, separate queues can be formed for each line, and a single-subscripted set would be defined and called, say, QUEUE. There would be as many sets as lines, and reference to the queue for the i*th* line would be made with the notation QUEUE(I). For sets having more than one subscript, a compound entity is defined, naming the set and the two or more permanent entities on which it is based. For example, EVERY HUSBAND AND WIFE OWNS A FAMILY defines an array called FAMILY that has the size of HUSBAND times WIFE, assuming both HUSBAND and WIFE have been defined as permanent entities.

The sets are organized as list structures, which will be described more fully in the next chapter. This means that every individual temporary entity that might belong to the set has two words put aside as pointers, identifying its predecessor and successor in the set. In addition, it has another word indicating whether it is, in fact, in the set or not.

The set's owner, either the system or a permanent entity, also has three items associated with the set. Two are pointers identifying the first and last member of the set, and the third is a counter showing how many members are in the set. All these special items associated with a set member or owner are established automatically by the program. They each have a name created from the set name by prefixing a single character as follows:[1]

Pointers Associated with			
Set Member		Set Owner	
P.*name*	Predecessor	F.*name*	First member
S.*name*	Successor	L.*name*	Last member
M.*name*	Membership	N.*name*	Counter

12-2

Set Organization

Sets can be organized as FIFO, LIFO, or ranked sets. A FIFO (first-in, first-out) discipline means that entities are kept in the order of the time at which they join the set. A new arrival becomes the last set member, and if the system is told to take the first member, it takes the one that has been in the set longest. A LIFO (last-in, last-out) discipline is the exact opposite: new arrivals go to the front of

---

[1]Some pointers are redundant; for example, a predecessor pointer is not needed for a FIFO set. Other pointers may not be needed in particular applications, such as the first or last pointers of a ranked set. The user can suppress any unwanted pointers.

the set, and a request to take the first will select the one that last arrived in the set. Sets with these two disciplines are defined in the preamble with a statement in the form:

DEFINE set AS A FIFO (or LIFO) SET

If a set is to be ranked, the preamble must contain a statement that identifies the attribute of the member entities on which the ranking is based, and tells whether the order is for increasing or decreasing values of the attribute. The statement takes the following form:

DEFINE set AS A SET RANKED BY HIGH (or LOW) attribute

If the word HIGH (and LOW) is omitted, the system assumes the ranking to be by high values. If the system is not told otherwise, members of the set that have the same value of ranking attribute will be filed in the order in which they join the set. It is possible, however, to add a control phrase specifying that some other attribute of the member entities be used as a secondary criterion for ranking. The phrase added to the definition statement says, "THEN BY second attribute." The phrase must be separated by a comma.

12-3

Set Controls

All sets are initially empty. Their membership increases or decreases through the use of FILE and REMOVE statements, respectively. The simplest form of the FILE statement names the entity to be filed and the set in which it is to be filed, as follows:

FILE entity IN set

The entity will be filed as the last member to join the set, which places the entity record according to the discipline defined for the set.

The statement can specifically say "FIRST IN set" or "LAST IN set" which uses the owner pointers to file the entity record as requested, irrespective of the set discipline. It is also possible to file an entity at an interior point of a set, using a FILE statement which has one of the following two forms:

FILE name AFTER (or BEFORE) name

The first name identifies the entity being filed, and the second name identifies the set member to which it is to be adjacent. (This form of FILE cannot be used with a ranked set.)

The REMOVE statement can correspondingly remove the first or last member of a set, using one of the following forms:

REMOVE FIRST (*or* LAST) *variable* FROM *set*

Note carefully, the *variable* appearing in this statement is not identifying the member to be removed: it identifies a location which is to hold a pointer identifying the member that is removed. The REMOVE statement can also take a specific form, as follows:

REMOVE *name* FROM *set*

It is illegal to attempt removing an entity from an empty set. The content of a set can be checked by examining the counter associated with the set. It is more common to use a form of the IF statement, as follows:

IF *set* IS (*or* IS NOT) EMPTY, *statement*

The *statement* can be any SIMSCRIPT statement, or group of statements, that will be executed if the stated condition is true (or not true). Control otherwise passes to the next statement in the program that follows the ELSE or ALWAYS matching the IF statement.

The most common method of referencing entities in sets is with the FOR phrase, which is a looping control phrase. For example

```
FOR  EACH name IN set,
   DO
   . . .
   REMOVE  FIRST name FROM set
   . . .
   LOOP
```

If the set happens to be empty, this loop is not executed.

12-4

Telephone System Model 2

As an example of how sets are programmed, the telephone system will be simulated for the case in which calls that find all links busy wait for one to become free. (Busy calls, however, are still lost immediately.) We define a set, called QUEUE, to hold the waiting calls. It is an unsubscripted set which belongs to the system.

Figure 12-1 shows the preamble for the new model, most of which is the same as for the first model. Since calls can now wait, one of the statistics to be derived

```
 1   ''    TELEPHONE SYSTEM - MODEL 2
 2   PREAMBLE
 3   NORMALLY,MODE IS INTEGER
 4   EVENT NOTICES INCLUDE ARRIVAL AND CLOSING
 5        EVERY DISCONNECT HAS A DIS.CALL
 6   TEMPORARY ENTITIES
 7        EVERY CALL HAS AN ORIGIN AND A DESTINATION
 8             AND AN ATIME
 9             AND MAY BELONG TO THE QUEUE
10   DEFINE LINKS.IN.USE, MAX.LINKS, BLOCKED, BUSY, FINISHED,
11        STOP.TIME AND NOWAIT AS VARIABLES
12   DEFINE INTER.ARRIVAL.TIME, MEAN.LENGTH AND ATIME AS REAL VARIABLES
13   PERMANENT ENTITIES
14        EVERY TLINE HAS A STATE
15   THE SYSTEM OWNS THE QUEUE
16   DEFINE QUEUE AS A FIFO SET
17   DEFINE QTIME AS A REAL VARIABLE
18   ACCUMULATE LQ AS THE AVERAGE OF N.QUEUE
19   TALLY WQ AS THE AVERAGE OF QTIME
20        DEFINE SECONDS TO MEAN UNITS
21        DEFINE MINUTES TO MEAN * 60 UNITS
22   END
```

Figure 12-1. Telephone system - model 2, preamble.

from the simulation will be the mean time calls wait. For this purpose, the temporary entities representing the calls need another attribute in which to record the time at which waiting begins. The attribute will be called ATIME. Line 8 of the preamble adds this to the definition of CALL. The next line records the fact that a CALL record can belong to the queue. Upon reading this line, the system will set aside space for the three member attributes described above. (Lines 7, 8, and 9 are actually one statement.) Another statistic to be added to the earlier model is a count of how many calls have to wait, and a system variable in which to build up the count has been added in line 11. Line 12 has also been extended to note that ATIME is a real variable, since the SIMSCRIPT clock time is recorded in that mode. The queue itself is defined in line 15, which also specifies that the queue is owned by the system. Line 16 defines the queue discipline as first-in, first-out. Lines 17, 18, and 19 are concerned with gathering statistics, and they will be explained later.

The MAIN program for model 2 is shown in Fig. 12-2. It is the same as the MAIN program for model 1, except that the output report generator has been extended to print more statistics covering information about calls that wait to be connected.

Figure 12-3 shows the new form of the routine programming the arrival of a call. The test, at line 2, is extended to prevent an endless loop if too many lines become engaged. It allows for the fact that a call waiting for a link will tie up the calling line. Because the program is now to compute waiting times, line 4 has been added to place the current clock time, at time of creation, in the ATIME attribute of the call.

If the created call can be immediately connected, lines 12 and 14 note that the

```
 1   MAIN
 2   READ N.TLINE, MAX.LINKS, INTER.ARRIVAL.TIME, MEAN.LENGTH AND STOP.TIME
 3   CREATE EVERY TLINE
 4   IF N.TLINE < 2*MAX.LINKS + 2
 5        PRINT 1 LINE THUS
  TOO FEW LINES SPECIFIED. SIMULATION ABANDONED.
 6        STOP
 7   ELSE
 8   SCHEDULE AN ARRIVAL NOW
 9   SCHEDULE A CLOSING IN STOP.TIME MINUTES
10   START SIMULATION
11
12   ''   REPORT TO BE WRITTEN AT END OF SIMULATION
13
14            PRINT 1 LINE THUS
SIMULATION OF TELEPHONE SYSTEM - MODEL 2
15        SKIP 3 OUTPUT LINES
16        PRINT 5 LINES WITH N.TLINE, MAX.LINKS, INTER.ARRIVAL.TIME,
17             MEAN.LENGTH, STOP.TIME THUS
  NUMBER OF LINES              *
  NUMBER OF LINKS              *
  MEAN INTER-ARRIVAL TIME      *.* /SECONDS
  MEAN CALL LENGTH             *.* /SECONDS
  SIMULATED TIME               * /MINUTES
18        SKIP 2 OUTPUT LINES
19        PRINT 4 LINES WITH FINISHED, BUSY, BLOCKED, NOWAIT THUS
  CALLS PROCESSED              *
  BUSY CALLS                   *
  BLOCKED CALLS                *
  IMMEDIATE CONNECTIONS        *
20   SKIP 1 OUTPUT LINE
21   PRINT 2 LINES WITH WQ THUS
  AVERAGE WAIT TIME FOR ALL CALLS
  CONNECTED.                   *.* /SECONDS
22   SKIP 1 OUTPUT LINE
23   PRINT 2 LINES WITH LQ THUS
  AVERAGE QUEUE LENGTH OF CALLS THAT WAITED,
  WHETHER CONNECTED OR NOT.    *.**
24        STOP
25   END
```

Figure 12-2. Telephone system - model 2, MAIN routine.

waiting time is zero and add one to the count of calls that did not have to wait. If the call is blocked, line 22 now makes the calling line busy, and line 23 enters the call in the set called QUEUE. Calls that find the called party busy are treated as before: two counters are incremented and the call is destroyed.

The new routine for disconnecting a call, shown in Fig. 12-4, is longer than that previous because, when a call is disconnected and there is a call waiting, the link that has become free must be offered to the waiting call. It is possible that the call getting the offer finds the called party is busy, in which case that call is destroyed, and the offer is repeated to any other call that may also happen to be waiting, until either a connection is made or there are no more waiting calls. A DO loop, under the control of an UNTIL statement, is used for this purpose.

There is no need to change the closing event: it remains the same as that shown in Fig. 11-7. Figure 12-5 shows the output report produced for model 2 when run for the same conditions as used for the model 1 run.

```
1   EVENT ARRIVAL
2        IF N.TLINE >= 2*MAX.LINKS + N.QUEUE + 2,
3            CREATE A CALL
4            LET ATIME(CALL) = TIME.V
5            LET ORIGIN(CALL) = RANDI.F(1,N.TLINE,1)
6            UNTIL STATE(ORIGIN(CALL)) = 0,
7                LET ORIGIN(CALL) = RANDI.F(1,N.TLINE,1)
8            LET DESTINATION(CALL) = RANDI.F(1,N.TLINE,1)
9            WHILE DESTINATION(CALL) = ORIGIN(CALL),
10               LET DESTINATION(CALL) = RANDI.F(1,N.TLINE,1)
11           IF STATE(DESTINATION(CALL)) = 0 AND LINKS.IN.USE < MAX.LINKS,
12               LET QTIME = 0
13               ADD 1 TO LINKS.IN.USE
14               ADD 1 TO NOWAIT
15               LET STATE(ORIGIN(CALL)) = 1
16               LET STATE(DESTINATION(CALL)) = 1
17               SCHEDULE A DISCONNECT GIVEN CALL
18                   IN EXPONENTIAL.F(MEAN.LENGTH,1) SECONDS
19           ELSE
20               IF LINKS.IN.USE = MAX.LINKS,
21                   ADD 1 TO BLOCKED
22                   LET STATE(ORIGIN(CALL)) = 1
23                   FILE CALL IN QUEUE
24               ELSE
25                   ADD 1 TO BUSY
26                   ADD 1 TO FINISHED
27                   DESTROY CALL
28               ALWAYS
29           ALWAYS
30       ALWAYS
31       SCHEDULE AN ARRIVAL IN EXPONENTIAL.F(INTER.ARRIVAL.TIME,1) SECONDS
32       RETURN
33   END
```

Figure 12-3. Telephone system - model 2, arrival event.

```
1    EVENT DISCONNECT GIVEN CALL
2         LET STATE(ORIGIN(CALL)) = 0
3         LET STATE(DESTINATION(CALL)) = 0
4         DESTROY CALL
5         SUBTRACT 1 FROM LINKS.IN.USE
6         ADD 1 TO FINISHED
7         UNTIL QUEUE IS EMPTY OR LINKS.IN.USE = MAX.LINKS,
8             DO
9             REMOVE FIRST CALL FROM QUEUE
10            IF STATE(DESTINATION(CALL)) NE 0,
11                ADD 1 TO BUSY
12                ADD 1 TO FINISHED
13                LET STATE(ORIGIN(CALL)) = 0
14                DESTROY CALL
15            ELSE
16                ADD 1 TO LINKS.IN.USE
17                LET STATE(DESTINATION(CALL)) = 1
18                LET QTIME = TIME.V - ATIME(CALL)
19                SCHEDULE A DISCONNECT GIVEN CALL
20                    IN EXPONENTIAL.F(MEAN.LENGTH,1) SECONDS
21            ALWAYS
22            LOOP
23        RETURN
24   END
```

Figure 12-4. Telephone system - model 2, disconnect event.

```
             SIMULATION OF TELEPHONE SYSTEM - MODEL 2


                   NUMBER OF LINES                50
                   NUMBER OF LINKS                10
                   MEAN INTER-ARRIVAL TIME        12.0 /SECONDS
                   MEAN CALL LENGTH              120.0 /SECONDS
                   SIMULATED TIME                 200 /MINUTES


                   CALLS PROCESSED               979
                   BUSY CALLS                    267
                   BLOCKED CALLS                 123
                   IMMEDIATE CONNECTIONS         642

                   AVERAGE WAIT TIME FOR ALL CALLS
                   CONNECTED.                     2.4 /SECONDS

                   AVERAGE QUEUE LENGTH OF CALLS THAT WAITED,
                   WHETHER CONNECTED OR NOT.      .24
```

Figure 12-5. Telephone system - model 2, output report.

12-5

Gathering Statistics in SIMSCRIPT

The COMPUTE statement is used to evaluate an expression, and, if it is carried over a range of values, it can be expanded to create such statistics as the mean or standard deviation of the computed values. When the quantity to be monitored is an unsubscripted system variable, or any attribute of a permanent or temporary entity, the system can be instructed to collect certain statistics automatically by statements in the preamble. If the attribute of a permanent entity is specified, an array is set aside by the system for gathering the statistics for each member of the permanent entity type. When the attribute of a temporary entity is specified, each entity record that is created has space reserved for the statistics.

One statistical statement is the TALLY statement. It has a counting action, which means it records whenever the value of the variable it is monitoring is set, irrespective of the times at which the values are set. The other statement, ACCU-MULATE, has an integrating action, meaning that it weights the current value by the length of time the variable has had that value, before it changes the variable to its new value. The TALLY statement produces the type of statistics that were called counts and summary statistics in Sec. 8-8. The ACCUMULATE statement produces the type of statistics that were called utilizations and occupancies in Sec. 8-9, and that were illustrated in Figs. 8-11 and 8-12. Both statements name the quantity that is to be monitored, and give a list describing the statistics that are to be gathered and the names to be used for the results. The statistics that can be gathered are indicated in the following list, which gives the terms by which they are to be described:

Statistic	Alternative or Abbreviation
NUMBER	NUM
SUM	—
MEAN	AVERAGE, AVG
SUM.OF.SQUARES	SSQ
MEAN.SQUARE	MSQ
VARIANCE	VAR
STD.DEV	STD
MAXIMUM	MAX
MINIMUM	MIN

The term STD.DEV means, of course, the standard deviation. The other quantities are apparent from their names, except perhaps NUMBER: this is the number of times values of the variable were collected.

Lines 18 and 19 of the preamble, shown in Fig. 12-1, are examples of statistical commands. The variable LQ is being defined as the mean value over simulated time of the queue length, and the variable WQ is being defined as the mean time calls spend waiting to be connected (which includes zeros for calls that do not wait.) To compute the wait time, another system variable, QTIME, has been defined in line 17. The two examples of Fig. 12-1 have only one listed item, but the lists can be extended. If the standard deviation of the waiting time were wanted, in addition to the mean value, line 19 could be written as follows:

```
TALLY WQ AS THE AVG AND SQ AS THE STD OF QTIME
```

12-6
Searching Arrays

It is often necessary to search an array looking for the first example that has a given set of conditions. A search can be established with a FOR statement, which has the effect of setting up a DO loop. A FIND phrase can be added to record the value of some expression when the search first encounters a member of the array that meets the conditions. Usually the expression that is calculated involves the member that has been located: frequently, it will simply be the identification number of the member. A compound statement of the following general form is used:

> FOR *variable.1, control phrases,* = *expression.1* TO
> *expression.2* BY *expression.3,* FIND *variable.2* =
> THE FIRST *expression.4,* IF NONE, *any statement,*
> ELSE

Although written here in several lines, the entire statement is a single statement that can be written continuously. Typically, the FOR clause takes the form FOR I = 1 TO N, so that the array is searched from the first to the N*th* member. The FIND statement places in *variable.2* the value of *expression.4* when the first member satisfying the control phrases is met. The words THE FIRST are optional. If no member meeting the conditions is found, the program executes the statement after the IF NONE phrase, which will usually be a GO TO statement. The program otherwise goes to the statement following the ELSE phrase. If the only purpose of the search is to identify the member, then a simplified form "FIND THE FIRST CASE" is used. In any case, *variable.1* is the appropriate index number.

The control phrases can be WITH, UNLESS, WHILE, or UNTIL statements in which two expressions are compared, using the following general format:

> WITH *expression.1 comparison expression.3*

The comparisons that can be made, and the codes used are

GR	Greater than
GE	Greater than or equal
EQ	Equal to
NE	Not equal to
LS	Less than
LE	Less than or equal to

The conventional mathematical symbols may also be used to indicate the comparison.

As an example, suppose the array STATE for the telephone lines, instead of just holding a 1 when the line is busy, holds a code number identifying the caller. To search the lines to see whether a particular caller is active, the following command could be given:

> FOR LINE = 1 TO N.TLINE, WITH STATE(LINE) = CALLER,
> FIND THE FIRST CASE, IF NONE, GO TO NOT.ACTIVE, ELSE

The statement will leave in the location LINE the number of the line that has its entry, for STATE, set to the number placed in CALLER, if such a line exists. Otherwise the program transfers to NOT.ACTIVE.

In this example there was only a need to identify the line. If the purpose were to make some calculation involving the line (using other attributes that have not been defined here), the phrase THE FIRST CASE would be replaced by an arithmetic expression making the calculation—if necessary, using LINE as an index. The result would be stored in a variable named by the arithmetic expression.

Because the array in this case is part of the definition of a permanent entity, the FOR statement can be written in another form. It can say, simply, FOR EACH LINE.

12-7

Searching Sets

In the case of searching a set, the members are not in an array that can be scanned by an index. The FOR phrase then takes the form

FOR EACH *variable* OF *set*

As the program scans the set, the *variable* of this phrase will pass through the pointers of the set, following the order defined by the set discipline. The particular values assumed are not usually meaningful to the user, but the program needs the definition of a variable to establish control over the set. For example, to find the first call that has waited longer than 10 seconds, the following statement could be used:

FOR EACH MEMBER OF QUEUE, WITH TIME.V − ATIME(MEMBER) > 10,
FIND THE FIRST CASE, IF NONE, GO TO PASS, ELSE

The statement will search all waiting calls looking for one that has a waiting time greater than ten seconds. Because a pointer to the call is all that is needed, the phrase closes with "THE FIRST CASE," instead of specifying an expression which is to be evaluated.

Exercises

In exercises 12-1 through 12-4, modify the telephone system simulation to introduce the following changes:

12-1 Compute the average waiting time of all calls that have to wait, whether they are connected or not.

12-2   Compute the average and the maximum queue length.

12-3   Assign the length of the call as an attribute and organize the queue so that the call that will finish first is connected first.

12-4   Assume calls to line number 1 wait if the line is busy, keeping a link engaged. Organize the queue of calls for line 1 on a FIFO discipline.

12-5   Program the manufacturing shop model 2 of Fig. 9-5 in SIMSCRIPT.

12-6   Reprogram the bus service problem of exercise 11-6 assuming that passengers are to queue with a FIFO discipline and that if they are not able to board one bus, they wait for the next bus.

13

SIMULATION PROGRAMMING TECHNIQUES

13-1
Entity Types

The entities of a system that need to be modeled in a simulation fall into two categories, forming a distinction which is particularly significant when designing simulation programs. Consider the telephone system, which we have now seen programmed in GPSS and SIMSCRIPT. The number of telephone lines was fixed for a given run. The number of calls, however, was not fixed. Throughout the simulation, calls were being created and destroyed.

Entities such as the lines, which remain fixed in number, are called *permanent entities*. Entities such as the calls, which are transient, are called *temporary entities*. (We have seen that SIMSCRIPT uses those specific terms. We will shortly be discussing their interpretation in GPSS.)

The significance of the difference is brought out when planning the data structures of the program. Being fixed in number, and, usually having the same data elements to describe the attributes of each member of the set, the permanent entities are easily represented in the form of fixed tables. Sometimes the amount of data to be kept for individual members of a set of permanent entities is sufficiently variable to justify a data structure in which there is a fixed format section, arranged in tabular form, with some overflow segments that are used as needed.

Permanent entities can, therefore, be represented as indexed variables. Each member of a permanent entity set can be identified by number, and normal indexing techniques can be used to scan the range of entities. For example, in the telephone system simulation considered in Chap. 8, there described as a numerical computation, a fixed-size table was established to represent the telephone lines. (See Fig. 8-2.)

In that simulation, however, the calls in progress were represented by a variable length list. Each call had a line of data in the list, but its position in the list kept changing as calls arrived and left, or were re-sorted to maintain chronological order.

When we discussed delayed calls, in Sec. 8-5, another list of calls was introduced, the list of delayed calls. (See Fig. 8-6.) A call could not be on both lists simultaneously, and, at some point, a call was transferred from the delayed list to the calls-in-progress list. The records kept for a call were different on the two lists. Calls on the delayed call list did not have a completion time. They could not, therefore, be ordered chronologically by time of completion, as was done for the calls-in-progress list. In fact, no comment was made about the order in which delayed calls should be kept. It was implied that they were kept in order of arrival, but that is not the order in which they leave the list. The order of departure is essentially unpredictable.

13-2

List Processing

An effective programming technique for handling records such as the call records (which fluctuate in number, may need to be re-sorted, or may need to be moved from one category to another) is called *list processing*, (15). The records are then said to be chained together or in a *list*. The records of the entities in the list are identified in the computer memory by an address. If the records have more than one word, the address is assigned to one of the words, such as the first. One word, or field in a word, called a *pointer*, is set aside in each record for the purpose of constructing the list. In addition, a special word called the *list header* is provided for entering the list.

The records are chained together as illustrated in Fig. 13-1. The list header contains the address of the first record in the list. The pointer of the first record contains the address of the second record, and so on down the list. The last record in the list contains a special end-of-chain symbol in the pointer space to indicate that it is the last member. If the list happens to be empty, the list header contains the end-of-chain symbol.

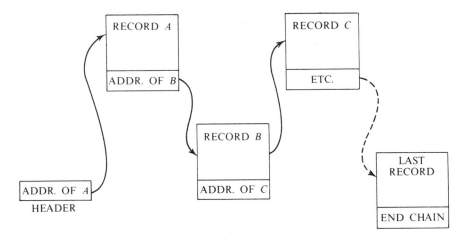

Figure 13-1. List structure.

Beginning from the header, the program is able to move down the list by following the chain of pointers. If the program needs to remove a record from the list, say record *B* from the list *ABC* . . . , it simply changes the pointer in *A* to point to *C*, as illustrated in Fig. 13-2. Correspondingly, to insert a record into the list, for example, to put *Z* in the list *ABC* . . . between *B* and *C*, the pointer *B* is set to *Z* and the pointer of *Z* is set to *C*. Re-sorting records is achieved by a series of removals and insertions.

With a first-in, first-out rule of ordering the records, it is convenient also to keep a *list trailer* that has the address of the last record, because new additions are

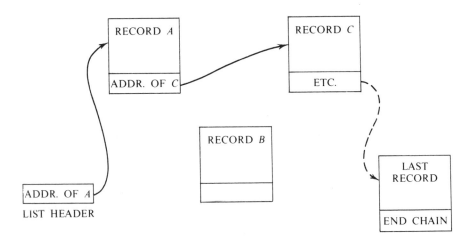

Figure 13-2. Removing a record from a list.

made at the end. The trailer record avoids the necessity of working along the list to find the last entry. The trailer will also contain the end-of-chain symbol when the list is empty.

It is sometimes necessary to scan the list from either end, in which case a second pointer is added to the records and the trailer record becomes the start for a search in the reverse direction, as shown in Fig. 13-3. The two types of pointer are called forward and backward pointers.

Other, more complicated organizations are possible to allow one record to become a member of more than one list. For example, suppose, in the telephone

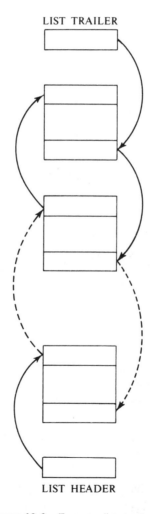

LIST TRAILER

LIST HEADER

Figure 13-3. Two-way list structure.

system, calls can originate from lines that are currently busy, and the system must keep track of calls delayed because they want to use a line either as a destination or an origin. The arrangement illustrated in Fig. 13-4 can then be used. Each line still has one header, but each call record has space for two forward pointers. A chain is formed for each line to hold waiting calls, in the order of requesting service. Because there are two pointers in a call record, one record can be on two lists, which would mean both the origin and destination of the call are busy.

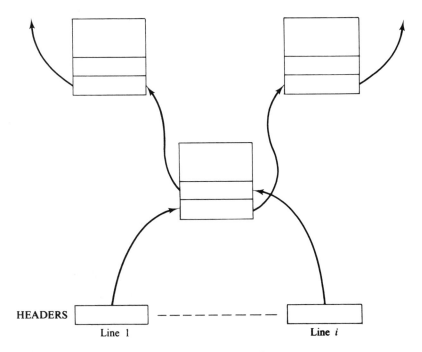

HEADERS

Line 1   Line *i*

Figure 13-4. Threaded lists.

The two pointers in the call record are not necessarily dedicated to constructing an origin or destination list. It is assumed that the program will be able to detect and use empty pointer space as it is needed. If a second header is added to each line, the header and forward pointers *can* be dedicated to forming separate chains for calls that want a line as an origin or destination: a more efficient organization if one type of delayed call is to be given priority over the other. Organizations that allow one record to be on more than one list simultaneously are said to form *threaded lists*, (16).

It is also possible to have a *sublist*, which is a list that has its header (or trailer) embedded in a member of some other list. All members of the sublist are members of the other list, and they can be reached through their header within that list, but they have some other common characteristic that justifies their being grouped together. For example, if telephone calls wait when either the destination or origin are busy, an alternative organization to the threaded lists would be to form a list of calls waiting for a line as a destination. A sublist could be formed of all calls waiting for the same destination and coming from the same origin.

There could, in fact, be many sublists of calls waiting for the same destination, as illustrated in Fig. 13-5. A header for all calls waiting to be connected to, say, line 7 points to calls organized in ascending order of the origin line number. Sub-

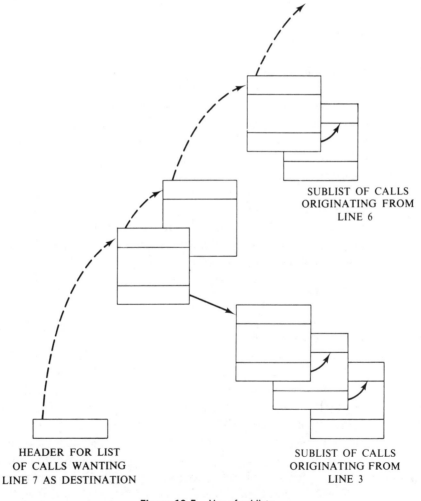

SUBLIST OF CALLS
ORIGINATING FROM
LINE 6

HEADER FOR LIST
OF CALLS WANTING
LINE 7 AS DESTINATION

SUBLIST OF CALLS
ORIGINATING FROM
LINE 3

Figure 13-5. Use of sublists.

lists are formed of all calls from the same origin, so that the main list has no more than one call from a given origin. Each of those calls, however, is potentially the beginning of a sublist: the figure shows sublists for lines 3 and 6.

Assume that, when a line becomes free, priority for its next use goes to calls needing it as a destination, and that those calls are given priority in the order of their origin number. When line 7 becomes free, the program uses the main chain to offer the line to calls waiting for that destination. If it reaches the calls from line 3 and finds line 3 is available as an origin, the program connects the first call of the sublist and moves the header of the sublist into the second call. If it finds that line 3 is not available, it moves to the next call of the main list, without wasting time testing each call trying to get to line 7 from line 3.

List processing involves a certain amount of "housekeeping" activity, but this can be organized in the form of general-purpose subroutines applicable to all lists. The operations of attaching and detaching records, and following the chains of pointers fall into this category. In particular, when the model calls for a record to be destroyed, the program must return the storage space for reuse. The record is sometimes put on a list of available space to be used again for the same purpose, or it may be returned to a general pool of available space. In the latter case, the fact that the records are scattered over the computer storage means that the fragments of unused space need to be merged together. Programs for performing this function are usually called garbage collectors, (3) and (17).

Most discrete system simulation languages are implemented by programming systems that have list processing capabilities. The technique, however, was not developed especially for simulation. Its origin lies in the development of programming systems for such nonnumeric applications as machine translation of languages, and theorem proving, (12) and (14).

13-3

Data Structures in SIMSCRIPT

As we have seen in Chaps. 11 and 12, SIMSCRIPT requires that the user specify each type of permanent and temporary entity, and these specific terms are used to identify the types. Event notices serve a special function, rather than representing entities of the system, but they are organized like temporary entities. From the specifications of the permanent entity types, together with the number of each type —which must be given as part of the program initialization—the program can construct fixed-size tables for each attribute of the permanent entities.

The structure of temporary entities is also given by the user. Records for these are created as the simulation proceeds, and they must be kept stored in list structures. If a temporary entity can belong to a set, that fact is declared in the preamble of the program, so that the program knows what pointer spaces must be included in the record. The definition of the set identifies the set owner, so that the program is also able to set aside space for header and trailer pointers. In addition, the system has an automatically defined set in which to file event notices in chronological order.

Most data structures needed for gathering statistics and carrying out computations are defined by the user, either as single system variables or as arrays. Some are automatically generated as a result of requests for standard statistics made in the preamble.

13-4
Data Structures in GPSS

GPSS does not use the terms permanent and temporary to describe entities. In the GPSS world-view, permanent entities are represented as facilities, storages, or logic switches. GPSS has prearranged tables for the information kept about facilities, storages, and logic switches. The format of the information is also prearranged. Even the maximum number of each type of item is prearranged, although the user can rearrange the numbers to adjust the allocation of space.

Other concepts of GPSS are more properly regarded as part of the statistics and computation facilities of the program. These are the functions, tables, savevalues, and variables. Tables of fixed size and given format are also used for these items. In the case of functions, tables, and variables, however, the amount of information is very variable. The programming technique that is used is to have a fixed size record for information that is always needed, and blocks of space allocated from a common pool to carry the variable information. Details of the variable space are carried in the fixed part of the record.

In GPSS, queues are counted in the category of statistical elements. Although the normal connotation of the word queue is of a fluctuating number of entities, a queue in GPSS is no more than a counter. The only information kept on queues is the count, of how many entities are in the queue, and statistics derived from the count, such as the maximum and mean length. In contrast to SIMSCRIPT, where a queue is usually defined as a set, and records are filed in the set, GPSS does not keep a record of which transactions are in a queue, just as it does not keep a record of which transactions are in a storage. If such records are needed, a user chain, or group, must be used. SIMSCRIPT, of course, can be programmed to define a queue in the GPSS sense.

Temporary entities in GPSS are represented by transactions, created at GEN-ERATE blocks. The user does not have to specify any details about the data structure of transactions unless the default values are unsatisfactory. The number of parameters, representing attributes of the temporary entities, and their data formats can, then, be reassigned by coding given at the GENERATE block.

Three types of sets are made available to the user: user chains, groups, and assembly sets. These are organized as lists, although the user is not required to make any specifications about the lists, other than providing information about the order of ranking for user chains if a discipline different from FIFO is needed.

The transactions are not only records of temporary entities, they are also used by the GPSS program to control the execution of events. The system has two major lists for this purpose: the future events chain, and the current events chain. In addition, there are many chains for holding transactions that are currently blocked. For each permanent entity represented by a facility, storage, or logic switch, GPSS automatically establishes lists for holding transactions that are waiting for that entity to be in a particular state. There are six lists for each facility or storage, and two for each logic switch. There are, therefore, fourteen conditions in which a transaction will join one, and only one, of these lists. They are:

Facility in use	Storage not empty
Facility not in use	Storage full
Facility interrupted	Storage not full
Facility not interrupted	Storage available
Facility available	Storage not available
Facility not available	Logic switch set
Storage empty	Logic switch not set

Transactions go on one of these lists when their movement is blocked—for example, when trying to enter a SEIZE block for a facility that is in use; or when trying to enter a GATE block, coded for a particular condition that does not currently exist.

These lists associated with the permanent entities are generically called *delay chains*. The user is not involved in the specification or manipulation of any of these lists. The way GPSS uses them in executing the simulation algorithm will be described shortly.

13-5

Implementation of Activities

An activity has been characterized as a process causing a change in the state of the system, that is, a change in one or more of the state descriptors that form the system image. The approaches taken in implementing activities in SIMSCRIPT and GPSS are strikingly different.

In SIMSCRIPT, activities are represented by event routines. As we saw in the application of SIMSCRIPT to the telephone system example, an event routine is a subroutine written by the user. All the direct changes implied by the activity are made by the event routine. At the same time, any statistics that will need to be reported are collected. There is considerable freedom in the way the event routine is organized, and the user is helped by many programming statements designed specifically for simulation. The resultant subroutines are compiled into machine code, giving the opportunity to produce efficient, fast code.

In GPSS, the changes are caused by macros, representing the block types. The user does not write code in the usual sense of the word.[1] The block diagram drawn by the user provides the parameters needed to implement the macros associated with the block types, and all this block diagram information is stored by the GPSS program.

The execution of the activities is carried out interpretively. While the code for the macros has been written and compiled in an efficient manner, the interpretive mode of operation involves overhead processing and prevents improving efficiency by combining the functions of blocks into more efficient code. However, the interpretive approach has the advantage of expediency, allowing models to be quickly assembled and debugged.

13-6

Simultaneous Events

In simulation programs, attention must be paid to conditions that exist when two or more events are scheduled to occur simultaneously. While this means that the events occur at the same instant in the system, the simulation program is forced to execute the events in sequence. There may be priorities assigned as attributes of the entities, in which case, the program should consider events in order of priority.

If no particular rule of ordering is made, the sequence of execution is, in effect, unpredictable. The choice of order may not be significant in the representation of the system. However, in trying to follow the detailed performance of the model, particularly when debugging the model, it may be difficult to interpret the simulated action. A study of the event history, with an understanding of how the simulation algorithm operates, may enable the order to be determined. However, as a practical matter, the precise sequence of events may be so involved that it is virtually impossible to untangle. For this reason, priority is sometimes used to control the ordering of simultaneous events, even though there may not be any corresponding distinction in the system.

---

[1]It is possible to introduce code written by the user, through the use of a block type called HELP.

The events that occur simultaneously may not be independent of each other. For example, two entities may be competing for the same item of equipment. The system can behave very differently according to the choice of order. Two simulation runs which logically appear to be exactly the same can give very different results. Some slight difference in the way the programs are assembled or executed can cause simultaneous events to be presented for execution in a different order. In particular, when random numbers are being used, different sequences of numbers may make significant differences between otherwise identical runs. Slight differences in the rounding-off of floating-point numbers, particularly when the clock time is recorded in that form, can have the same effect.

Another aspect of simultaneous events that must be considered is the fact that an entity moving through the system may be involved in activities that require no time. There can, in fact, be a string of such "zero-time" activities. The execution of an event involving a zero-time activity immediately produces another event due for execution at the same clock time. By itself, this may present no difficulty; but, if there are several different entities, all due to participate in simultaneous events, and any of them encounters zero-time activities, the question of choosing the order of the scanning can be important.

The condition is illustrated in Fig. 13-6, which represents a set of entities, $E_1, E_2, \ldots, E_n$, all due to begin moving at the *same* clock time and each headed for individual strings of zero-time activities of various lengths. The activities to be executed during this clock instant for the ith entity are $A_{ij}(j = 1, 2, \ldots, k_i)$, so that potentially there is a total of $\sum_{i=1}^{n} k_i$ events due for execution at the same instant.

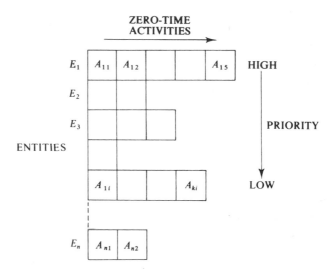

Figure 13-6. Representation of simultaneous events.

If no control is exercised over the order in which events are scanned, the order of execution will be unpredictable, leading to possible confusion and violation of priority rules.

In practice, the simulation program is not likely to be aware of all the possible simultaneous events. The fact that an entity is going into a zero-time activity is not usually realized until the activity is implemented, and so the existence of the second and higher columns of Fig. 13-6 is not known. In fact, the implementation of the program may be such that no attention is paid to any event other than the highest priority event of the currently posted events, that is, the top event in the first column of Fig. 13-6.

Two other factors need to be considered. The execution of an event may change the priority of an activity or an entity (including the entity causing the event). The other possibility is that an event may create new entities, or implement new activities, that are to be involved in events to occur at the current clock time. The point at which the new events or changes of priority will take effect, in relation to existing concurrent events, needs to be considered.

Different event scanning methods may cause differences in the order in which simultaneous events will be executed. It is important to understand the method of scanning (and generating) events in systems where simultaneous events can occur. In particular, it is important to understand whether any external factors, such as order of assembling a program, or machine dependencies, can cause changes.

13-7

Conditional Events

When an event cannot be executed at its scheduled time, the record for the event must lie dormant until the condition preventing its execution changes. It is then called a *conditional event*. The records of the conditional events are often kept in a separate list called a *ready* list, and the simplest programming arrangement is to rescan the ready list whenever the state of the system changes, to determine whether conditions have become favorable for any of the conditional events. There can be a cascading effect, with the execution of one event releasing one or more events at the same clock time, which in turn may release others, and so on, leading again to the problem of considering the order of processing simultaneous events.

To describe the problem involved, consider the situation shown in Fig. 13-7, which represents a ready list of conditional event records. The records are arranged in order of priority and are scanned from the highest to the lowest priority (top to bottom). When the scan of the list is started, it may read a record, say *C*, which is

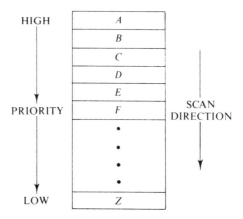

Figure 13-7. Scanning conditional events.

found to represent an event that has become released. This release may, in turn, allow the events represented by records B and F to be released. Record B has been passed by the scan and as a result, if the scan continues down the ready list, the lower priority record, F, may take advantage of the change, ahead of B. For complete accuracy, the ready list scan should be restarted whenever any change in the list occurs, and the process repeated with each detected change until the scan manages to get through the entire list without any further change.

13-8
Event Scanning

Most discrete system simulation languages are implemented by programming systems that use the "next event" scheduling method; that is, the clock time is advanced to the time of the next event, or next potential event. Some programs are able to mix discrete and continuous simulation, and they can be switched between a next-event and a uniform-step timing method. It is possible to use a uniform step method in a discrete system simulation, and there are circumstances under which such an approach can be more efficient, in terms of execution time, (13). If time is expressed in integral units only, it is sometimes more efficient to collect records of events in groups—according to the instant at which they occur—and advance through the lists, without specifically referring to clock time. Simulating logic circuits is an application in which that technique can be used, (18). Some programs do not concentrate on time but on the logical conditions under which an activity becomes eligible for execution, (20). They are effective for models in which there is a preponderance of conditional events.

The next-event method is based on the premise that most events have a known time at which they are due to occur. The beginning of an activity in the system will schedule a future event, which is the ending of the activity. List processing provides a convenient way of organizing records of future events by keeping them in chronological order. Usually a simple list is used, but experiments with more complex organizations have shown improvements in execution times, (6), (21), and (22).

In comparing different simulation languages, three general approaches to sequencing events have been proposed as a basis for classification, (10). These are called: event scheduling, process interaction, and activity scanning. The first two employ a next-event method of organizing event notices. The principal difference among them is in the scanning of simultaneous events. We have seen, in the last two sections, that simultaneous events can arise from the coincidence of events being scheduled at the same time, or from the release of conditional events. Whatever event scanning method is used, all simultaneous events that are capable of being executed should be executed. The different approaches can group the simultaneous events in slightly different ways, and, as a result, may produce different results if there is some interaction between the simultaneous events.

The *event scheduling* approach sees a system as a collection of overlapping activities. The beginning and ending of each activity are each regarded as separate events which are independently scheduled. A conditional event can be treated as a sub-event within the event routine that causes its release, or it can be scheduled as a separate, concurrent event. Similarly, if an entity is created and is to be involved in an immediate event, that event might be a sub-event or a separately scheduled event.

The *process interaction* approach concentrates on the individual entities, (5). The system is seen as a set of overlapping activities, causing events as they start and finish, but the activities form related groups, which are the processes. Once committed to a process, an entity will generally proceed through all the activities of the process. If the end of one activity implies the start of another for the same entity, these two events will be executed in sequence, and not scheduled separately. Similarly, if a non-zero activity is encountered, so that the start of an activity implies its immediate end, those two events will also be executed in sequence. An entity will, therefore, be taken through as many events of a process as presently possible. In particular, any conditional events released as the process is being executed will not normally be considered until the process has been carried as far as possible.

The *activity scanning*[2] approach does not specifically use the next-event method, although the simulation proceeds in uneven steps through successive events. All

---

[2]The term *machine-based* was originally used for this approach, (19). In contrast, the term *material-based* was used to describe the other two approaches that have been defined.

activities have a statement giving the conditions under which they may be started, including a specification of what entities and resources must be available, (9). Each active entity has an associated clock giving the time when the entity will end the activity in which it is engaged. Scanning the clocks determines which event occurs next. Following the change of state that occurs, all activities are scanned to see which can then be started. The clocks do not, in fact, need to show absolute time: they can be timers showing the length of the interval to the end of the activity, and which can be reset whenever the system moves to a new event.

The original classification of languages by Kiviat (10) has been updated by Fishman [(4), p. 85]. It includes several languages of historical interest which are now little used or obsolete. The most prominent event-scheduling languages are SIMSCRIPT and GASP. Among the process-interaction languages, GPSS and SIMULA are the most prominent. The best known examples of the activity-scanning approach are CSL (1) and its extension, ECSL (2), together with SIMON, (8).

Recently, SIMSCRIPT II.5 has been extended by the addition of concepts called *resources* and *activities* which allow the program to be used in a process-interaction manner. This is an alternative to the traditional event-scheduling mode of operation for SIMSCRIPT.

13-9

Execution of Simulation Algorithm
in SIMSCRIPT

We are now in a position to summarize the way the simulation algorithms are executed in both SIMSCRIPT and GPSS. We begin with SIMSCRIPT. [See Chap. 5 of (11).] It will be recalled that the principal steps involved in a simulation algorithm were identified, in Sec. 8-6, as being:

1. Find the next potential event.
2. Select an activity.
3. Test if the event can be executed.
4. Change the system image.
5. Gather statistics.

Performance of the first two steps in SIMSCRIPT is illustrated in Fig. 13-8. Event notices are the program's way of recording future events. The notices are filed in a list in chronological order. When the processing of all current events is complete, the program finds the next potential event by taking the event notice at the head of the list. By convention, the event notice is given the same name as the event routine it is due to schedule. The program selects the next activity by transferring to

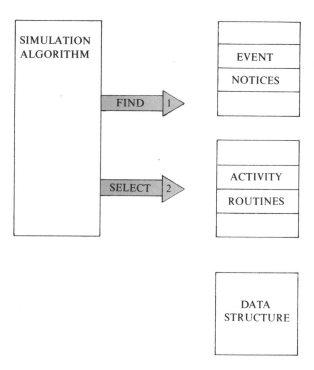

Figure 13-8. Execution of simulation algorithm in SIMSCRIPT-1.

that routine. These first two steps, therefore, are executed automatically by the SIMSCRIPT timing routine.

The other steps are executed in the individual event routines, programmed by the user. Figure 13-9 is a flowchart showing the general organization of an event routine. Unless it is known that the same event will be rescheduled immediately, the event notice is destroyed, since its purpose of transferring programming control has been accomplished, and the space it occupies should be recovered. Conditions to be met before the event can be executed are then tested. This is step three of the algorithm. If the conditions are not met and the event is not to be abandoned, it becomes a conditional event. The user must make some note of the conditional event, such as filing a record in a set, and make arrangements for the event to be executed when the necessary conditions come into effect.

Following a successful test, step four of the algorithm is executed by making the changes to the system image corresponding to the event. These changes, of course, depend upon the way the user has chosen to establish a system image. The fifth step, gathering statistics, is then performed. If the execution of the event is to lead to some other event, the flow of control is advanced by creating an event notice for the follow-on event, and scheduling its execution.

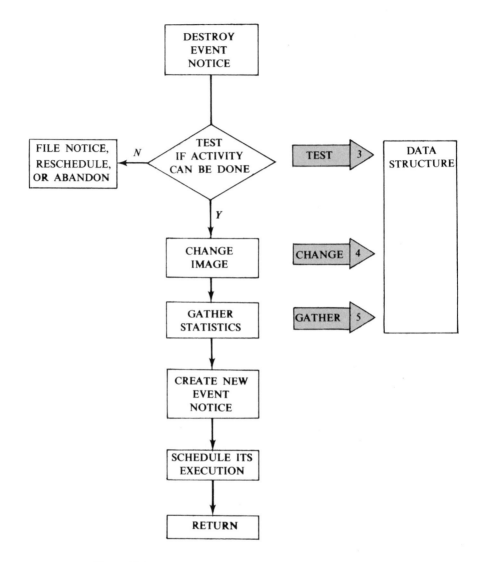

Figure 13-9. Execution of simulation algorithm in SIMSCRIPT-2.

With regard to simultaneous events, the event notices can be assigned priority, according to their type. It is also possible to make tie-breaking statements, for the event notice types, that will further order the notices on any sequence of attribute values, in either an ascending or descending order.

As has been seen, the user is responsible for activating conditional events when they become unblocked. This includes recognizing the relative priorities among the conditional events, and the priorities with respect to any other simultaneous events that may be represented by other event notices, due to be executed at the current time, but not yet examined.

The set manipulation capabilities of SIMSCRIPT allow records representing conditional events, of the same type, to be kept in priority order with the same capabilities as apply to the event notices. Maintaining relative priorities between conditional events of different types, or between conditional events and simultaneous events represented by event notices, needs some consideration. If there is some critical interaction between two or more types of conditional events, it may be necessary to break event routines into smaller units, so that event notices can be scheduled for immediate execution on behalf of a released conditional event. The priority control exercised by the system can then take over. For example, in model 2 of the telephone system, the connection of a call that has been waiting for a link to become free is made in the event routine that disconnects a call. The connection of a delayed call is a conditional event, and it is being treated as a sub-event of the disconnect routine. Similarly, the connection of a new arrival, when it can be made immediately, is being treated as a sub-event of the arrival routine. An alternative way of programming the model would be to treat the connection and disconnection of calls as separate event routines (with the arrival as a third). When the disconnect routine finds a call to be waiting, it can then schedule a connection event on behalf of the blocked call, to be executed at the current clock time. If there happened to be simultaneous arrivals at that moment, the relative priorities of the delayed call and the arrivals that have not yet been tested, will be correctly handled. As programmed in Chap. 12, the model is capable of giving a link to a delayed call that is of lower priority than a new arrival.

13-10

Execution of Simulation Algorithm in GPSS

Figure 13-10 will be used to illustrate how GPSS executes the simulation algorithm. [See Appendix 6 of (7).] Transactions in GPSS, while representing temporary entities, are also used by the program to schedule events. They can do so because each transaction carries a record of the time it is next due to move. The record is maintained automatically by the program. All transactions scheduled to move at some time ahead of the current clock time are filed on a future events chain, in chronological order and by priority for coincident times. While on the future events chain, therefore, transactions are similar to the event notices of SIMSCRIPT. When all the events at one clock time have been processed, the program finds the

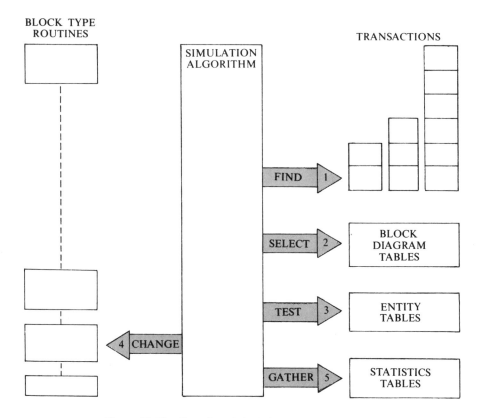

Figure 13-10. Execution of simulation algorithm in GPSS.

next potential event by taking the first transaction from the future events chain. The current block location of the transaction is also kept in the transaction record. Since the block diagram information is stored, the program can determine which is the next block, and so execute the second step of the simulation algorithm.

Information about the status of all facilities, storages, and logic switches is automatically maintained by the program. Knowing the next block, the program can determine whether conditions for entering the block are satisfactory, thereby executing step three of the algorithm. The block routines automatically record the changes to the system image (step four). The statistics for step five are also gathered automatically, although the user has some control over the statistics that are gathered, by the disposition of the block types associated with statistics.

As can be seen, all steps involved in the execution of the simulation are performed automatically. This includes the action of scheduling future events, since the **GENERATE** and **ADVANCE** blocks, which are the only block types that explicitly invoke time, include the actions of filing a transaction in the future events chain.

In addition to the future events chain, GPSS has a chain, called the current events chain, of transactions that are unable to move because of some blocking condition. The chain is, in effect, a ready list of conditional events. In Sec. 13-4, when discussing the data structure of GPSS, it was stated that a blocked transaction is placed on one of 14 types of delay chain, according to the condition causing the blocking. The current events chain and the delay chains operate together to form threaded lists of transactions. Physically, there is only one record of a transaction. When the transaction is found to be blocked, it is placed on the current events chain. Pointers in the transactions also connect together all transactions waiting for any of the conditions listed in Sec. 13-4, and tie the list of such transactions to a header associated with the particular facility, storage, or logic switch causing the blocking.

The flowchart of Fig. 13-11 describes how the event scanning of GPSS resolves simultaneous events. When the program has moved all the transactions that can move at one time, it updates the clock to the time of the first transaction on the future events chain. Since the chain is in chronological order, the clock is set to the time of the next potential event. The program transfers that transaction to the current events chain. It also moves any other transactions on the future events chain due to move at the same time. The transactions joining the current events chain are merged according to priority, which is the ordering rule for that chain. This ensures that, in the subsequent movement of transactions, the relative priorities of any conditional events, represented by transactions that were left on the current events chain, and the simultaneous events, represented by the transactions just transferred, are properly recognized.

The program then begins scanning the transactions on the current events chain to see if they can be moved. While on the current events chain, transactions are in one of two states: they are either active or inactive. If inactive, the transaction is also on a delay chain, and so it is known that it cannot move until there is a change in the state of some permanent entity. When that change occurs the program will automatically reset the transaction to active. In the meantime, the scan will bypass inactive transactions.

Since GPSS is organized according to the process interaction method, if it finds it can move a transaction, it will move the transaction through as many blocks as possible at the current time. If the transaction moves into an ADVANCE block (with a non-zero action time) it comes to represent a future event, and it is placed back on the future events chain. If the transaction becomes blocked, or it is found that an active transaction cannot, in fact, move, the transaction represents

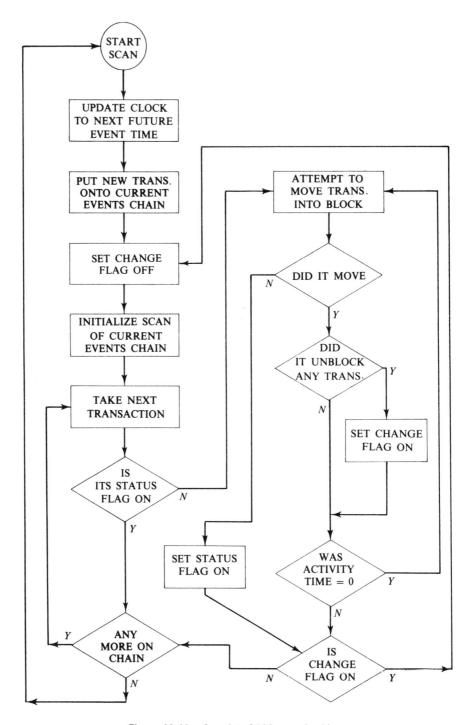

Figure 13-11. Complete GPSS scan algorithm.

a conditional event. It remains on the current events chain, and it is put into an inactive state.[3]

In the course of moving a transaction into a block that changes the state of any permanent entity, the block routine responsible for making the change checks the associated delay chains. If it finds transactions that have been waiting for that change, it removes them from the delay chain, and sets their status to active. When the current transaction has stopped moving, the program checks a change flag which will have been turned on if any of the transaction moves caused a change in the status of a permanent entity. If the flag is on, the program restarts the scan to ensure that any transaction of lower priority, released by one that just moved, is processed. The scanning continues until either there are no more transactions on the current events chain, or the program has been able to pass through the entire chain without the need for a rescan.

Bibliography

1   BUXTON, J. N., "Writing Simulations in CSL," *Comput. J.*, IX, no. 2 (1966), 137–143.

2   CLEMENTSON, A. T., "Extended Control and Simulation Language," *Comput. J.*, IX, no. 3 (1966), 215–220.

3   DEUTSCH, L. PETER, AND DANIEL G. BOBROW, "An Efficient, Incremental, Automatic Garbage Collector," *Commun. ACM*, XIX, no. 9 (1976), 522–526.

4   FISHMAN, GEORGE S., *Concepts and Methods in Discrete Event Digital Simulation*, New York: John Wiley & Sons Inc., 1973.

5   FRANTA, W. R., *A Process View of Simulation*, New York: American Elsevier Publishing Company Inc., 1977.

6   GONNET, GASTON H., "Heaps Applied to Event Driven Mechanisms," *Commun. ACM*, XIX, no. 7 (1976), 417–418.

7   GORDON, GEOFFREY, *The Application of GPSS V to Discrete System Simulation*, Englewood Cliffs, N. J.: Prentice-Hall Inc., 1975.

8   HILLS, P. R., edited by S. H. HOLLINGDALE, *SIMON — A Computer Simulation Language in ALGOL, in Digital Simulation in Operations Research*, pp. 105–115, New York: American Elsevier Publishing Company, 1967.

9   HUTCHINSON, GEORGE K., "Introduction to the Use of Activity Cycles as a

---

[3]Some complex blocking conditions cannot be represented by a simple delay chain. When these occur, the transactions remain on the current events chain in an active state, so that they are rechecked on every scan.

Basis for System's Decomposition and Simulation," *Simuletter*, VII, no. 1 (1975), 15–20.

10 KIVIAT, P. J., *Digital Computer Simulation: Computer Programming Languages* The Rand Corporation, RM-5883-PR, Santa Monica, Cal. 1969.

11 ——, R. VILLANUEVA, AND H. M. MARKOWITZ, (Ed. E. C. RUSSELL), *SIMSCRIPT II.5 Programming Language*, CACI, Inc., Los Angeles, Cal., 1975.

12 MCCARTHY, JOHN, *LISP 1.5 Programmer's Manual*, Cambridge, Mass.: The MIT Press, 1962.

13 NANCE, RICHARD E., "On Time Flow Mechanisms for Discrete System Simulation," *Manage. Sci.: Theory*, XVIII, no. 1 (1971) 59–73.

14 NEWELL, A., AND F. M. TONGE, "An Introduction to Information Processing Language V," *Commun. ACM*, III, no. 4 (1960), 205–211.

15 NICHOLLS, JOHN E., *The Structure and Design of Programming Languages*, Chap. 8, Reading, Mass.: Addison Wesley Publishing Company, 1975.

16 PERLIS, A. J., AND CHARLES THORNTON, "Symbol Manipulation by Threaded Lists," *Commun. ACM*, III, no. 4 (1960), 195–204.

17 SCHORR, H., AND W. M. WAITE, "An Efficient Machine-Independent Procedure for Garbage Collection in Various List Structures," *Commun. ACM*, X, no. 8 (1967), 501–506.

18 SZYGENDA, STEPHEN, CLIFF W. HEMMING, AND JOHN M. HEMPHILL, "Time Flow Mechanism for Use in a Digital Logic Simulation," *Proc. 1971 Winter Simulation Conf.*, WSC/SIGSIM, 562 Croydon Rd., Elmont, N.Y., 1974, 488–495.

19 TOCHER, K. D., "Some Techniques of Model Building," *Proc. IBM Scientific Computing Symposium on Simulation Models and Gaming*, IBM Corp., White Plains, N.Y., 1966, 119–155.

20 VAUCHER, J. G., "A 'Wait Until' Algorithm for General Purpose Simulation," *1973 Winter Simulation Conf. Proc.*, WSC/SIGSIM, Elmont, N.Y., 1973, 77–83.

21 ——, AND PIERRE DUVAL, "A Comparison of Simulation Event List Algorithms," *Commun. ACM*, XVIII, no. 4 (1975), 223–230. [See also *Corrigendum*, XVIII, no. 8 (1975), 462.]

22 WYMAN, F. PAUL, "Improved Event-Scanning Mechanism for Discrete Event Simulation," *Commun. ACM*, XVIII, no. 6 (1975), 350–353.

14

ANALYSIS
OF SIMULATION OUTPUT

14-1

Nature of the Problem

Once a stochastic variable has been introduced into a simulation model, almost all the system variables describing the system's behavior also become stochastic, because of the way endogenous events make one variable depend upon another. The values of most, if not all, of the system variables will fluctuate as the simulation proceeds, so that no one measurement can be arbitrarily taken to represent the value of a variable. Instead, many observations of the variable's value must be made, in order to make a statistical estimate of its true value. Some statement must also be made about the probability of the true value falling within a given interval about the estimated value. Such a statement defines a *confidence interval*. Without it, simulation results are of little value to a system analyst.

A large body of statistical methods has been developed over the years to analyze results in science, engineering, and other fields where experimental observations are made. Because of the experimental nature of system simulation, it seems natural to attempt applying these methods to simulation results. Unfortunately, most of them presuppose that the observations being made are mutually independent—a reasonable assumption when an experiment is being repeated, or independent samples are being selected.

Simulation results, however, are not likely to be mutually independent. A single simulation run will produce many "observed" values of a variable, but the value observed at one time is likely to be influenced by the value at some earlier time. For example, in the simulation of a waiting line, the time one entity spends waiting depends upon the number of entities that happened to be on the waiting line at the time it arrived. The need to account for the resultant interactions requires careful application of the established statistical methods. This need has also lead to the development of new statistical methods, and it is the subject of much current research work.

One concern of this newly developing statistical methodology is to ensure that the statistical estimates are consistent, meaning that, as the sample size increases, the estimate tends to the true value. Another concern is to control bias in measures of both mean values and variances. Bias causes the distributions of estimates based on finite samples to differ significantly from the true population statistics, even though the estimates may be consistent. A third, practical aspect of current research work is the attempt to develop sequential testing methods that will allow automatic controls to determine how long a simulation should be run in order to obtain a given level of confidence in its results, (9) and (11).

The purpose here is to review the problems involved, and to describe some of the methods being used to analyze simulation results. A more comprehensive discussion of the statistical aspects of simulation will be found in Ref. (3), (4), (5), (6), (10), and (14).

14-2
Estimation Methods

We first review some of the statistical methods commonly used to estimate parameters from observations on random variables. Usually, a random variable is drawn from an infinite population that has a stationary probability distribution with a finite mean, μ, and finite variance, σ^2. This means that the population distribution is not affected by the number of samples already made, nor does it change with time. If, further, the value of one sample is not affected in any way by the value of any other sample, the random variables are mutually independent. Random variables that meet all these conditions are said to be independently and identically distributed, usually abbreviated to i.i.d. Under broad conditions that can be expected to hold for simulation data, the *central limit theorem* can be applied to i.i.d. data. The theorem states that the sum of n i.i.d. variables, drawn from a population that has a mean of μ and a variance of σ^2, is approximately distributed as a normal variable with a mean of $n\mu$ and a variance of $n\sigma^2$.

As was described in Sec. 7-9, any normal distribution can be transformed into a standard normal distribution, that has a mean of 0 and a variance of 1. Let

$x_i(i = 1, 2, \ldots, n)$ be the n i.i.d. random variables. Using the central limit theorem, and applying the transformation, gives the following (approximate) normal variate:

$$z = \frac{\sum\limits_{i=1}^{n} x_i - n\mu}{\sqrt{n}\,\sigma}$$

Dividing top and bottom of this fraction by n, and defining $\bar{x} = \dfrac{1}{n}\sum\limits_{i=1}^{n} x_i$, we have

$$z = \frac{\bar{x} - \mu}{\sigma/\sqrt{n}}$$

The variable $\bar{x}$ is the *sample mean*. It can be shown to be a consistent estimator for the mean of the population from which the sample is drawn. Since the sample mean is the sum of random variables, it is itself a random variable. As a result, a confidence interval about its computed value, needs to be established.

The probability density function of the standard normal variate is illustrated in Fig. 14-1. The integral from $-\infty$ to a value u is the probability that z is less than or equal to u. The integral is usually denoted by $\Phi(u)$ and tables of its value are widely available. Suppose the value of u is chosen so that $\Phi(u) = 1 - \alpha/2$, where α is some constant less than 1, and denote this value of u by $u_{\alpha/2}$. The probability that z is greater than $u_{\alpha/2}$ is then $\alpha/2$. The normal distribution is symmetric about its mean, so the probability that z is less than $-u_{\alpha/2}$ is also $\alpha/2$. Consequently, the probability

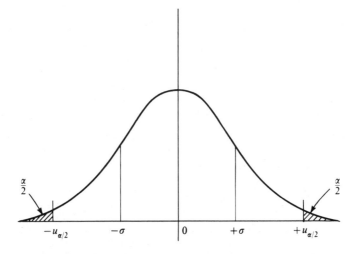

Figure 14-1. Probability density function of standard normal variate.

that z lies between $-u_{\alpha/2}$ and $u_{\alpha/2}$ is $1 - \alpha$. That is,

$$\text{Prob}\left\{-u_{\alpha/2} \leq z \leq u_{\alpha/2}\right\} = 1 - \alpha$$

In terms of the sample mean, this probability statement can be written

$$\text{Prob}\left\{\bar{x} + \frac{\sigma}{\sqrt{n}} u_{\alpha/2} \geq \mu \geq \bar{x} - \frac{\sigma}{\sqrt{n}} u_{\alpha/2}\right\} = 1 - \alpha$$

The constant $1 - \alpha$ (usually expressed as a percentage) is the confidence level and the interval

$$\bar{x} \pm \frac{\sigma}{\sqrt{n}} u_{\alpha/2}$$

is the confidence interval. The size of the confidence interval depends upon the confidence level chosen. Typically, the confidence level might be 90%, in which case $u_{\alpha/2}$ is 1.65. The statement then says that μ will be covered by the confidence interval $\bar{x} \pm 1.65\sigma/\sqrt{n}$ with probability 0.9; meaning that, if the experiment is repeated many times, the confidence interval can be expected to cover the value μ on 90% of the repetitions.

In practice, the population variance σ^2 is not usually known; in which case, it is replaced by an estimate calculated from the formula

$$s^2 = \frac{1}{n-1} \sum_{i=1}^{n} (x_i - \bar{x})^2 \qquad (14\text{-}1)$$

The normalized random variable based on σ^2 is replaced by a normalized random variable based on s^2. This has a Student-t distribution, with $n - 1$ degrees of freedom. The quantity $u_{\alpha/2}$ used in the definition of a confidence interval given above, is replaced by a similar quantity, $t_{n-1,\alpha/2}$, based on the Student-t distribution, for which tables are also readily available.

The Student-t distribution is strictly accurate only when the population from which the samples are drawn is normally distributed. It is common practice, however, to rely upon this distribution when it is being assumed that the distribution is normal, as when the central limit theorem is being invoked.

Expressed in terms of the estimated variance, s^2, the confidence interval for $\bar{x}$ is defined by

$$\bar{x} \pm \frac{s}{\sqrt{n}} t_{n-1,\alpha/2} \qquad (14\text{-}2)$$

14-3

Simulation Run Statistics

In addition to the assumption of normality inherent in the use of the central limit theorem, the method of establishing confidence intervals, outlined in the previous section, is based on two other assumptions. It is assumed that the observations are mutually independent, and it is assumed that the distribution from which they are drawn is stationary. Unfortunately, many statistics of interest in a simulation do not meet these conditions. To illustrate the problems that arise in measuring statistics from simulation runs, a specific example will be discussed.

Consider a single-server system in which the arrivals occur with a Poisson distribution and the service time has an exponential distribution. The queuing discipline is first-in, first-out with no priority. Suppose the study objective is to measure the mean waiting time, defined as the time entities spend waiting to receive service and excluding the service time itself. The problem can be solved analytically. (The solution was given in Sec. 7-12.) This system is commonly denoted by M/M/1, which indicates; first, that the inter-arrival time is distributed exponentially; second, that the service time is distributed exponentially; and, third, that there is one server. (The M stands for Markovian, which implies an exponential distribution.)

In a simulation run, the simplest approach is to estimate the mean waiting time by accumulating the waiting time of n successive entities and dividing by n. This measure, the sample mean, is denoted by $\bar{x}(n)$ to emphasize the fact that its value depends upon the number of observations taken. If $x_i (i = 1, 2, \ldots, n)$ are the individual waiting times (including the value 0 for those entities that do not have to wait), then

$$\bar{x}(n) = \frac{1}{n} \sum_{i=1}^{n} x_i \qquad (14\text{-}3)$$

Waiting times measured this way are not independent. Whenever a waiting line forms, the waiting time of each entity on the line clearly depends upon the waiting times of its predecessors. Any series of data that has this property of having one value affect other values is said to be *autocorrelated*. The degree to which the data are autocorrelated can be measured in ways that will be briefly described in a later section.

Under broad conditions that can normally be expected to hold in a simulation run, the sample mean of autocorrelated data can be shown to approximate a normal distribution as the sample size increases (7). The usual formula for estimating the mean value of the distribution, Eq. (14-3), remains a satisfactory estimate for the mean of autocorrelated data. However, the variance of the auto-correlated data is not related to the population variance by the simple expression σ^2/n, as occurs for independent data. A term must be added to account for the autocorrelation. The term is usually positive in the case of the M/M/1 system, so that, if it is ignored, the variance is underestimated, but, in other systems, it can be negative, resulting in an overestimate.

Another problem that must be faced is that the distributions may not be stationary. In particular, a simulation run is started with the system in some initial state, frequently the idle state, in which no service is being given and no entities are waiting. The early arrivals then have a more than normal probability of obtaining service quickly, so a sample mean that includes the early arrivals will be biased. As the length of the simulation run is extended, and the sample size increases, the effect of the bias will die out. For a given sample size starting from a given initial condition, the sample mean distribution is stationary; but, if the distributions could be compared for different sample sizes, the distributions would be slightly different. The analytical solutions previously quoted are for the steady state values to which the distributions converge as the sample size increases.

Figure 14-2, which is adapted from Ref. (16), and is based on theoretical results derived in Ref. (13), shows how the expected value of the sample mean depends upon the sample length, for the M/M/1 system, starting from an initial empty state, with a server utilization of 0.9.[1] It is known that the steady state mean in this case is 8.1. It can be seen that the mean value is biased below the steady state value. The bias diminishes as the sample size increases but, even with a sample size of

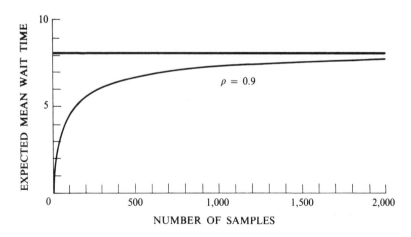

Figure 14-2. Mean wait time in M/M/1 system for different sample sizes.

[1]See Sec. 7-11 for a definition of the term *server utilization.*

2,000, the mean has still only reached about 95% of the steady state value. The steady state value will be approached more rapidly for lower levels of server utilization, but, unfortunately, high server utilization cases are usually the ones of interest in simulation studies.

14-4

Replication of Runs

One way of obtaining independent results is to repeat the simulation. Repeating the experiment with different random numbers for the same sample size n gives a set of independent determinations of the sample mean $\bar{x}(n)$. Even though the distribution of the sample mean depends upon the degree of autocorrelation, these independent determinations of the sample mean can properly be used to estimate the variance of the distribution. Suppose the experiment is repeated p times with independent random number series. Let x_{ij} be the ith observation in the jth run, and let the sample mean and variance for the jth run be denoted by $\bar{x}_j(n)$ and $s_j^2(n)$, respectively. For that jth run, the estimates are:

$$\bar{x}_j(n) = \frac{1}{n} \sum_{i=1}^{n} x_{ij} \qquad (14\text{-}4)$$

$$s_j^2 = \frac{1}{n-1} \sum_{i=1}^{n} [x_{ij} - \bar{x}_j(n)]^2$$

Combining the results of p independent measurements gives the following estimates for the mean waiting time, $\bar{x}$, and the variance, s^2, of the population:

$$\bar{x} = \frac{1}{p} \sum_{j=1}^{p} \bar{x}_j(n) \qquad (14\text{-}5)$$

$$s^2 = \frac{1}{p} \sum_{j=1}^{p} s_j^2(n) \qquad (14\text{-}6)$$

The value of $\bar{x}$ is an estimate for the mean waiting time, and s^2 can be used to establish a confidence interval, according to the expression of Eq. (14-2), with $p - 1$ degrees of freedom.

Figure 14-3 shows the result of applying the procedure to experimental results for the M/M/1 system. Results are shown for server utilizations of 0.1, 0.3, 0.4, 0.5, and 0.6. In each case, the experiment has been repeated from an initial idle state, different random numbers being used on each repetition. The results show the estimated mean waiting time calculated from Eq. (14-3) as a function of sample size, n. Measurements were made in steps of $n = 5$ for $\rho = 0.2, 0.3$, and 0.4, and

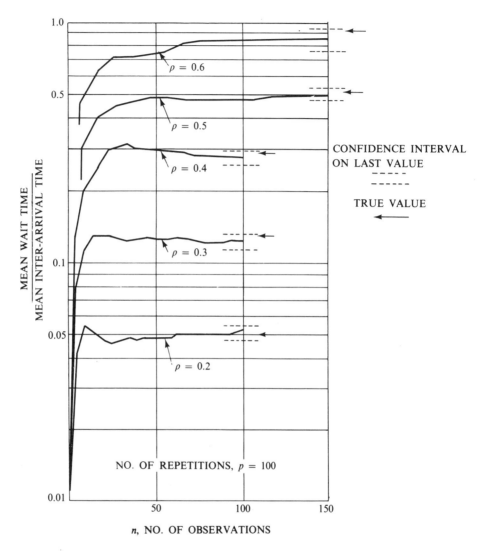

Figure 14-3. Experimentally measured wait time in M/M/1 system
for different sample sizes.

$n = 10$ for $p = 0.5$ and 0.6. Each case is for 100 repetitions ($p = 100$). Also shown are the 90% confidence intervals calculated for the highest value of n used in each case, and the known steady state values of the mean waiting times.

The p repetitions of n observations involve a total of $N = p.n$ observations. In a computer based simulation, the total time spent carrying out calculations will be roughly proportional to N. The question of how best to divide the N observations between the number of repetitions and the length of the individual runs has given rise to much discussion, (8). Increasing the number of repetitions decreases the size of the confidence interval, since the variance estimate is approximately inversely proportional to p. Normally, this is a desirable effect, since it reduces the range of uncertainty about the estimate of the mean. However, the desirability assumes the estimate of the mean is unbiased. A short confidence limit, centered on a biased estimate, can easily fail to cover the true value being estimated. In the replication of simulation runs, if the number of runs is increased at the cost of shortening the individual runs (in order to keep N constant), the estimate of the mean will be more biased, as a result of the initial empty state.

Law (16), for example, reports results of experiments with the M/M/1 system, at a server utilization of 0.9. The entire process of computing a confidence interval by replication of runs starting from an empty state, in this case at a 90% confidence level, was repeated 400 times for different combinations of n and p.[2] For each combination of n and p, the total number of observations used in each determination of a single confidence interval, N, was kept at 12,800.

At a 90% confidence level, it is to be expected that 90% of the 400 independently determined confidence intervals will cover the true mean. Taking the steady state mean as the true mean, the results showed that for 5 replications, each of 2,560 samples ($n = 2,560$ and $p = 5$), 83% of the confidence intervals covered the true mean. At the other extreme of the experiment, 40 replications, each of 320 samples ($n = 320$ and $p = 40$), the figure dropped to 9%. It can be estimated from Fig. 14-2 that, at a sample size of 2,560, the mean wait time is about 7.8, which is only 0.3 below the steady state mean. At a sample size of 320, the mean is about 6, which is 2.1 below the steady state mean. The shorter confidence interval, achieved with the greater number of replications, is bought at the cost of a much greater bias— resulting in the large difference in the accuracy.

From this evidence, we see that it is preferable to keep the number of repetitions as low as possible, bearing in mind the need to approximate a normal distribution with the sample means. These results, of course, are for a simple system. However, they are indicative of the type of research work being conducted in the study of the simulation process.

14-5

Elimination of Initial Bias

The experimental results given in Fig. 14-3, together with those reported in Ref. (1), clearly show the need to remove the intial bias, or reduce its effect. Two general approaches can be taken to remove the bias: the system can be started in

---

[2]The notations used in Ref. (16) for n, p, and N are m, k, and n, respectively.

a more representative state than the empty state, or the first part of the simulation run can be ignored.

In some simulation studies, particularly of existing systems, there may be information available on the expected conditions that makes it feasible to select better initial conditions. The ideal situation is to know the steady state distribution for the system, and select the initial condition from that distribution. In the study previously discussed, Law repeated the experiments on the M/M/1 system, supplying an initial waiting line for each run, selected at random from the known steady state distribution of the waiting line. The case of 40 repetitions of 320 samples, which previously resulted in a coverage of only 9%, was improved to a coverage of 88%. Of course, the theoretical knowledge on which this technique is based is not usually available. However, experience with an existing system, or similar type of system, could provide a reasonable approximation.

The more common approach to removing initial bias is to eliminate an initial section of the run. The run is started from an idle state and stopped after a certain period of time. The entities existing in the system at that time are left as they are. The run is then restarted with statistics being gathered from the point of restart. As a practical matter, it is usual to program the simulation so that statistics are gathered from the beginning, and simply wipe out the statistics gathered up to the point of restart. No simple rules can be given to decide how long an interval should be eliminated. It is advisable to use some pilot runs starting from the idle state to judge how long the initial bias remains. This can be done by plotting the measured statistic against run length as has been done in Fig. 14-3.

Another disadvantage of eliminating the first part of a simulation run is that the estimate of the variance, needed to establish a confidence limit, must be based on less information. The reduction in bias, therefore, is obtained at the price of increasing the confidence interval size.

14-6

Batch Means

Another approach to the problem of establishing confidence intervals for simulation results does not rely upon replication, but uses a single long run, preferably with the initial bias removed. The run is divided into a number of segments to separate the measurements into batches of equal size. The sample means are then treated as independently, identically distributed variables. A complete run consists of N observations that are broken into p batches of size n. (It is assumed that N is exactly divisible by p.) In effect, the experiment is equivalent to repeating an experiment of length n a total of p times, with the final state of one run becoming the initial condition for the next. Denoting the ith observation in the jth batch by x_{ij}, Eqs. (14-4) through (14-6) can be used to estimate the mean and standard deviation of the variable being measured; and Eq. (14-2) can then be used to establish a confidence interval, (with $p - 1$ degrees of freedom).

This way of repeating a run is preferable to starting each run from an initial idle state, since the state at the end of a batch is a more reasonable initial state than the idle state. However, the connection between the batches introduces correlation. Sometimes, the batches are separated by intervals in which measurements are discarded in order to eliminate the correlation. Clearly, this throws away useful information. Conway (2) demonstrated that the variance to be expected by using all the data and accepting the correlation between batches is less than that obtained from the reduced amount of data obtained by separating the batches. It seems to be preferable, therefore, to work with adjoining batches.

The batch mean method has the advantage of the repetition method without the necessity of eliminating the initial bias on each repetition. However, it is necessary to assume that the individual batch means are independent. The assumption can be justified if the batch length is sufficiently long. The effect of autocorrelation is that the value of one piece of data affects the value of following data. The effect usually diminishes as the separation between the data increases, and beyond some interval size it may reasonably be ignored. If the batch size is greater than this interval, the batch means may be treated as independent. It remains a matter of judgement to choose a suitable batch size. It might reasonably be speculated that the interval over which the batch is measured should be at least as great as the interval excluded from the beginning of a run to remove initial bias. If that value has been determined, it can also be used as a batch size. However, the only safe procedure is to use a test run in which to try a batch size and test for the presence of correlation in the results (2). Another approach is to repeat the calculations with several batch sizes and test for consistency of results. By making the batch sizes multiples of each other, it is possible to perform the operation in a single run.

An important practical aspect of the batch method is that it does not entail the simultaneous presence of all the data to carry out the calculations. The batch means can be calculated as the simulation run proceeds. Computer space is only required to accumulate the sum of the batch means and the sum of their squares, together with an accumulation of the numbers forming the current batch mean. The batch method, therefore, forms a good basis for sequential testing methods.

In discussing the replication method, it was pointed out that there is a trade-off between the number of repetitions and the run length. With the batch method there is a similar trade-off between batch size and number of batches. Since the number of batches corresponds to the number of samples of an assumed normal distribution, it is again advisable to hold this number to a reasonable limit to meet the assumption, and maximize the batch size, in order to reduce the correlation between batches.

The experiments reported in (16) tested the batch mean method against the replication method for many combinations of n and p. The results showed that, in almost all cases, the batch mean method was superior to the replication method, and that the difference was statistically significant. Again, the results were for simple systems, although, in this test, more complex systems than the M/M/1 system were tested.

14-7

Regenerative Techniques

We have seen that a discrete system is described by a number of state descriptors that are changing value at specific points in time. To know the state of the system (to the extent that it is modeled), it is necessary to know the values of the state descriptors. Suppose all the values are known for one point of time. Most discrete systems are such that exactly the same set of values will occur some time later. In fact, *that* set of values will continue to recur at certain random intervals of time.

A system with this property is said to be *regenerative*. A particular set of values is chosen as a reference; the times at which the same set of values recurs are called *regeneration points*. Between successive regeneration points, the system is said to execute a *tour*. The time the system spends in a tour is called an *epoch*, (15). Consider, again, measuring the waiting time in the M/M/1 system as an example. Suppose the system starts from the empty state, and measurements begin from the time the first entity arrives for service. We use the return to the empty state to define the regeneration points. At the time the measurements begin, the system becomes busy with the first entity. If a second entity arrives before the first completes service, the system will immediately begin serving the second entity when the first completes service. The server will then remain busy, and this may happen again with the following arrivals. Eventually, (assuming the server utilization is less than 1) there will be no other entity waiting when service finishes. The system then becomes idle, and remains so until the next arrival occurs. The point at which the system becomes busy again marks a regeneration point. This process is repeated as the system moves through successive cycles of a busy period followed by an idle period. Each of these cycles is a tour.

The number of entities served in a tour varies. It could happen that an entity finding the system idle, makes the system busy, but there is no other arrival by the time service for that entity finishes. The system will fall back to the idle state, and there will have been a tour involving only one entity (with a waiting time of zero).

As each busy period begins, the system regenerates itself. That is, its future behavior is independent of its past behavior. The system behavior in any tour is independent of its behavior in any other tour. As a result, samples of any statistic taken from the tours are i.i.d. In particular, the epochs are i.i.d. As a result, the behavior of the system in a tour is characteristic of steady state behavior, and the estimates that are based on the tour statistics are free of the initial bias that proved to be so troublesome with replicated tests.

Suppose there are p complete tours when measuring an M/M/1 system. Let Y_j be the sum of the waiting times for the entities served in the jth tour, and let n_j be the number of entities served in that tour. Because of the regeneration principle, both Y_j and n_j are i.i.d. samples, each from its own distribution. Note, however, that Y_j and n_j are not mutually independent: on the contrary, they are strongly, positively correlated, since a larger value of Y_j can be expected to occur with a larger value of n_j.

Disregarding the tours for the time being, suppose there is a total of N observations, and the individual waiting times are $w_1, w_2, \ldots, w_N$. Then, a consistent estimate of the mean waiting time, denoted by $\bar{W}$, is given by

$$\bar{W} = \frac{w_1 + w_2 + \cdots + w_N}{N}$$

If the individual waiting times are now grouped according to their tours, this expression may be written in the form

$$\bar{W} = \frac{Y_1 + Y_2 + \cdots + Y_p}{n_1 + n_2 + \cdots + n_p}$$

If, further, we define

$$\bar{Y} = \frac{1}{p} \sum_{j=1}^{p} Y_j \tag{14-7}$$

$$\bar{n} = \frac{1}{p} \sum_{j=1}^{p} n_j \tag{14-8}$$

the estimate of the mean waiting time is given by

$$\bar{W} = \frac{\bar{Y}}{\bar{n}} \tag{14-9}$$

The numbers $\bar{Y}$ and $\bar{n}$ are, of course, the estimates of the means for the sum of the waiting times in a tour, and the number of entities served in a tour. Equation (14-9) states that the estimate of the mean waiting time is the ratio of these two numbers. Using the notation $E(x)$ to denote the expected value of a random variable x, it also follows that

$$E(W) = \frac{E(Y)}{E(n)} \qquad (14\text{-}10)$$

To construct a confidence interval for W, it is necessary to estimate the variance of the statistic $\bar{W}$. To avoid dealing directly with the ratio of random variables, we define the following variable:

$$V_j = Y_j - E(W)n_j \qquad (14\text{-}11)$$

Note that $E(W)$, the expected value of the waiting time, is a constant (even though its value is not known). This makes V_j a linear combination of two random variables that are i.i.d. It follows that V_j is also i.i.d. If we define $\bar{V}$ as the sample mean of V_j, it can also be seen that

$$\bar{V} = \bar{Y} - E(W)\bar{n} \qquad (14\text{-}12)$$

In addition, we have

$$E(V) = E(Y) - E(W)E(n)$$

Substituting the value of $E(W)$ from Eq. (14-10), it follows that the expected value of the variable V is zero.

Since the values of V_j are i.i.d., the central limit theorem can be applied to their sum. Letting σ^2 be the variance of V, we construct the following standard normal variate (since the value of $E(V)$ is zero):

$$z = \frac{\bar{V}}{\sigma/\sqrt{p}}$$

This leads to the following probability statement, (using $u_{\alpha/2}$ and α as defined in Sec. 14-2):

$$\text{Prob}\left\{-u_{\alpha/2} \leq \frac{\bar{V}}{\sigma/\sqrt{p}} \leq u_{\alpha/2}\right\} = 1 - \alpha$$

Substituting for $\bar{V}$ from Eq. (14-12), and multiplying the expression by $\sigma/\sqrt{p}$, results in

$$\text{Prob}\left\{-\frac{\sigma u_{\alpha/2}}{\sqrt{p}} \leq \bar{Y} - E(W)\bar{n} \leq \frac{\sigma u_{\alpha/2}}{\sqrt{p}}\right\} = 1 - \alpha$$

This can be further rearranged to give

$$\text{Prob}\left\{\frac{\bar{Y}}{\bar{n}} + \frac{\sigma u_{\alpha/2}}{\bar{n}\sqrt{p}} \geq E(W) \geq \frac{\bar{Y}}{\bar{n}} - \frac{\sigma u_{\alpha/2}}{\bar{n}\sqrt{p}}\right\} = 1 - \alpha$$

We have, therefore, constructed a confidence interval about the expected mean wait time defined by the points

$$\bar{W} \pm \frac{\sigma u_{\alpha/2}}{\bar{n}\sqrt{p}} \tag{14-13}$$

The variance of V_j, σ^2, however, is not known, but it can be estimated from Eq. (14-11), which defines V_j in terms of Y_j and n_j. Since Y_j and n_j are not mutually independent, we must take note of their covariance. Let s_V^2 denote the sample variance of V, based on the p pairs of measurements taken from the tours; then it can be shown that

$$s_V^2 = s_{11}^2 - 2\bar{W}s_{12}^2 + \bar{W}^2 s_{22}^2 \tag{14-14}$$

where s_{11}^2, s_{22}^2, and s_{12}^2 are the sample variance of Y_j, the sample variance of n_j, and the sample covariance[3] of (Y_j, n_j), respectively. From the definition of V_j, given by Eq. (14-11), the expression for s_V^2 should involve $E(W)$. Since this is not known, it is replaced by its estimated value, $\bar{W}$. In terms of the observed data, the values of s_{11}^2, s_{22}^2, and s_{12}^2 are

$$s_{11}^2 = \frac{1}{p-1}\sum_{j=1}^{p}(Y_j - \bar{Y})^2 = \frac{1}{p-1}\sum_{j=1}^{p}Y_j^2 - \frac{p\bar{Y}^2}{p-1} \tag{14-15}$$

$$s_{22}^2 = \frac{1}{p-1}\sum_{j=1}^{p}(n_j - \bar{n})^2 = \frac{1}{p-1}\sum_{j=1}^{p}n_j^2 - \frac{p\bar{n}^2}{p-1} \tag{14-16}$$

$$s_{12}^2 = \frac{1}{p-1}\sum_{j=1}^{p}(Y_j - \bar{Y})(n_j - \bar{n}) = \frac{1}{p-1}\sum_{j=1}^{p}Y_jn_j - \frac{p\bar{Y}\bar{n}}{p-1} \tag{14-17}$$

Replacing the unknown variance of V by its estimate, the estimate of the mean waiting time, and its confidence interval are

$$\bar{W} = \frac{\bar{Y}}{\bar{n}} \tag{14-18}$$

$$\bar{W} \pm \frac{s_V u_{\alpha/2}}{\bar{n}\sqrt{p}} \tag{14-19}$$

---

[3]To avoid confusion, we adopt the usual convention of using a squared quantity to denote a variance estimate. Note, however, that s_{12}^2 can be negative.

Summarizing the application of the regeneration method, we see that the following steps are involved:

1) Gather simulation data for p tours
2) For each tour, j, record the total wait time, Y_j, and the number served, n_j
3) Compute the statistics $\bar{Y}, \bar{n}, s_{11}^2, s_{22}^2, s_{12}^2,$ and s_V^2 from Eqs. (14-7), (14-8), (14-15), (14-16), (14-17), and (14-14), respectively
4) Substitute in Eq. (14-18) for the estimate of the mean waiting time, and in Eq. (14-19) for a confidence interval, using α and $u_{\alpha/2}$ as defined in Sec. 14-2.

14-8

Time Series Analysis

The three methods for making estimates of the variance of the sample mean that have been described so far—the replication of runs, batch means, and the regenerative method—have been based on the principle of getting (or closely approximating) independent samples. Another approach is to accept the presence of autocorrelation, and to estimate the variance of the sample mean by methods that have been developed in the study of time series.

With such studies, it is not usually possible to rerun a stochastic process. Furthermore, the autocorrelation contains valuable information, since it describes the underlying nature of the process producing the time series. This has led to the development of techniques for measuring the autocorrelation. Given measures of the autocorrelation, it is possible to compensate for its presence, and deduce estimates of the variance of the population from which the observations are drawn.

Virtually all time series are based on data observed at uniform intervals of time. To apply the time series techniques to simulation results, therefore, a single run is made, with observations being made at uniform intervals of time. For convenience, suppose the observations are made at unit time intervals, and the run is for a time of T time units, so that there are T observations.

Autocorrelation is measured by a series of autocovariances that show the extent to which values separated by s time units affect each other. The autocovariance between values of a variable made at two different times, X_t and X_u, is defined as

$$R_{tu} = E[(X_t - \mu)(X_u - \mu)]$$

If we assume that the value of μ is independent of the indices t and u, and that R_{tu} depends only upon the separation $s = t - u$, the stochastic process is said to be *covariance stationary*. The autocovariance then exists for all integer values of t and u, but has the property of symmetry that $R_{tu} = R_{ut}$. Since the value then depends only on the separation, the autocovariance is usually denoted by R_s, and the property of symmetry says that $R_s = R_{-s}$. For convenience, in the following discussion, s is allowed to take all values $0, \pm 1, \pm 2, \ldots$, but the case of $s = 0$ is actually the definition of the variance of X_t.

Given a set of observations at T uniform intervals of time, the autocovariances are estimated from the equations

$$R_s = \frac{1}{T} \sum_{t=1}^{T-s} (X_t - \bar{X})(X_{t+s} - \bar{X}) \qquad (s = 0, 1, \ldots, T-1) \qquad (14\text{-}20)$$

$$\bar{X} = \sum_{t=1}^{T} X_t$$

The estimation of an unbiased sample mean, $V(\bar{X})$, from the autocovariance values involves some computational subtlety, but an acceptable formula [(10, pp. 281–284] is

$$V(\bar{X}) = \frac{T}{T-k} \left\{ R_0 + 2 \sum_{s=1}^{k-1} \left(1 - \frac{s}{T} \right) R_s \right\} \qquad (14\text{-}21)$$

The estimate is taken to have $3k/2n$ degrees of freedom.

It will be noticed that only the first $k - 1$ autocovariance estimates are involved. Normally, the values of the autocovariances decrease as s increases. The value used for k should be large enough to include the autocovariances of significant size, but, as s increases, the accuracy of the autocovariance estimate decreases. For a fixed number of observations, the number of pairs of points available for the estimate decreases as s increases, leading to difficulties in judging the accuracy of the sum of the autocovariances. One approach [(10), p. 285] is to carry out the estimation for different values of s, looking for the estimate to stabilize. An example discussed in Ref. (10), for the M/M/1 system at a server utilization of 0.9, required a k of about 300.

It is apparent that the method involves a considerable amount of calculation. It also suffers the disadvantage of requiring that all the data be simultaneously available—unlike the batch mean approach, which is able to summarize results as the run proceeds.

14-9

Spectral Analysis

The analysis of time series through autocorrelation is intimately related to another way of viewing time series. A time series may also be regarded as the summation of oscillations of different frequencies. The spectrum of frequencies and the amplitudes of the oscillations can be formally related to the autocovariance. Essentially the same calculations involved in the estimation of autocorrelation can be applied to derive a spectral analysis of a simulation run. The results can be used to estimate sample mean variances, (17).

A spectral analysis, however, can provide more information than is contained in the estimate of a mean value. Comparing two systems on the basis of mean values of such factors as waiting time or queue length is a rather gross comparison. Two systems may show no significant difference in their mean values, but their transient behavior may be significantly different. One system may respond slowly to deviations from its mean values while the other may respond rapidly. Depending upon the purpose of the system, one performance may be preferable to the other. A simple comparison of means will mask the difference but a spectral analysis can distinguish the difference by showing whether the spectrum emphasizes low or high frequencies.

Another interesting use of spectral analysis has been to test a simulation model by comparing the spectral pattern of the simulation output against the corresponding pattern derived from the system itself, (12). This, of course, can only be done when analyzing an existing system. However, in complex systems, where it is unlikely that a simulation could produce a precise event-by-event replication of the true system response, a reasonable match of the spectra can indicate that the system response is being reproduced correctly.

14-10

Autoregressive Processes

Among all possible stochastic processes, there is a subset, called *autoregressive processes*, having a structure that makes the evaluation of its autocorrelation properties relatively easy to calculate. The values of an autoregressive variable, X_t, are defined by the equation

$$X_t = -b_1 X_{t-1} - b_2 X_{t-2} - \cdots b_p X_p + \epsilon_t \qquad (14\text{-}22)$$

where ϵ_t is a random variable (usually assumed to be normally distributed), and the following conditions hold:

$$\mu = E(X_t)$$
$$E(\epsilon_t) = 0$$
$$E(\epsilon_t^2) = \sigma^2$$
$$E(\epsilon_s, X_t) = 0, \quad s > t$$
$$E(\epsilon_s, \epsilon_t) = 0, \quad s \neq t$$

The process is determined by p coefficients $b_i (i = 1, \ldots, p)$, together with the first $p - 1$ values of X_t, which must be given as initial conditions. Equation (14-22) states that the tth value of X_t $(t \geq p)$ is a linear sum of the previous $p - 1$ values, plus a random component, ϵ_t. The conditions state that the random variable, ϵ_t, has a zero mean, finite variance, and all its autocovariances are equal to zero. Further, the random component is not correlated with X_t. The definition can be written more compactly in the form:

$$\sum_{s=0}^{p} b_s (X_{t-s} - \mu) = \epsilon_t \qquad b_0 = 1$$

It can be shown that the autocovariances, which are still defined by Eq. (14-20), are related by the following set of equations, known as the *Yule-Walker* equations:

$$\sum_{s=0}^{p} b_s R_{s-r} = \begin{cases} \sigma^2 & r = 0 \\ 0 & r \neq 0 \end{cases}$$

This is a set of $p + 1$ linear equations involving the p coefficients $b_s (s = 1, \ldots, p; \ b_0 = 1)$, $2p + 1$ autocovariances, $R_s (s = -p, \ldots, -1, 0, 1, \ldots, p)$, and σ^2. However, the symmetry in the autocovariances means there are only $p + 1$ distinct autocovariances.

Assume that a series of observations, X_t, is generated by an autoregressive process of order p. Equation (14-20) provides estimates of the $p + 1$ autocovariances, $R_s, (s = 0, \ldots, p)$. Taking the p Yule-Walker equations that correspond to $r \neq 0$ gives a means of deriving estimates of the coefficients of the autoregressive process, $\hat{b}_s, (s = 1, \ldots, p)$. The Yule-Walker equation corresponding to $r = 0$ gives an estimate for σ^2 as follows:

$$\hat{\sigma}^2 = \sum_{s=0}^{p} \hat{b}_s \hat{R}_s \tag{14-23}$$

It can be shown [(10), pp. 286–287] that, by forming the quantity,

$$\hat{b} = 1 + \sum_{s=1}^{p} \hat{b}_s \tag{14-24}$$

an estimate of the variance of the sample mean, $V(X)$, is given by

$$m = \frac{\hat{\sigma}^2}{\hat{b}^2} \tag{14-25}$$

The degrees of freedom of the estimate are taken to be f, where

$$f = \frac{n\hat{b}}{(2p + 1)\hat{b} - 4 \sum_{s=0}^{p} s\hat{b}_s}$$

The question of what value to assume for p, the order of the autoregressive process, is left open. However, experiments with this approach have suggested that small values of p are adequate: certainly, the values needed for a good approximation appear to be much less than the large number of autocovariance calculations needed for the approach described in Sec. 14-8, where no assumptions were made about the nature of the stochastic process. In any event, it is possible to establish statistical tests for deciding whether to accept a particular value of p. There are also ways of computing the autocovariance estimates recursively, that is, deriving higher order values from lower order values. When combined, these techniques provide a means of applying sequential tests that can decide, automatically, how long a simulation should be run in order to measure a statistic to a given degree of confidence [(10), pp. 254–260].

Bibliography

1   CHENG, R. C. H., "Note on the Effect of Initial Conditions on a Simulation Run," *Oper. Res. Q.*, XXVII, no. 2 (1976), 467–470.

2   CONWAY, R. W., "Some Tactical Problems in Digital Simulation," *Manage. Sci., Theory*, X, no. 1 (1963), 47–61.

3   CRANE, MICHAEL J., AND DONALD L. IGLEHART, "Simulating Stable Stochastic Systems — I: General Multiserver Queues," *J. ACM.*, XXI, no. 2 (1974), 103–113.

4   ——, "Simulating Stable Stochastic Systems — II: Markov Chains," *J. ACM.*, XXI, no. 2 (1974), 114–123.

5   ——, "Simulating Stable Stochastic Systems — III: Regenerative Processes and Discrete-Event Simulations," *Oper. Res.*, XXIII, no. 1 (1975), 33–45.

6   ——, "Simulating Stable Stochastic Systems — IV: Approximation Techniques," *Manage. Sci., Theory*, XXI, no. 11 (1975), 1215–1225.

7   DIANANDA, P. H., "Some Probability Limit Theorems With Statistical Applications," *Proc. Cambridge Philos. Soc.*, XLIX, no. 3 (1953), 239–246.

8   FISHMAN, GEORGE S., "The Allocation of Computer Time in Comparing Simulation Experiments," *Oper. Res.*, XVI, no. 2 (1968), 280–295.

9   ——, "Estimating Sample Size in Computing Simulation Experiments," *Manage. Sci., Theory*, XVIII, no. 1 (1971), 21–38.

10 ——, *Concepts and Methods in Discrete Event Digital Simulation*, New York: John Wiley & Sons, Inc., 1973.

11 ——, "Achieving Specific Accuracy in Simulation Output Analysis," *Commun. ACM*, XX, no. 5 (1977), 310–315.

12 FITZSIMMONS, JAMES A., "The Use of Spectral Analysis to Validate Planning Models," *Socio.-Econ. Plan. Sci.*, VIII, no. 3 (1974), 123–128.

13 HEATHCOTE, C. R., AND P. WINER, "An Approximation for the Moments of Waiting Times," *Oper. Res.* XVII, no. 1 (1969), 175–186.

14 KLEINJNEN, JACK P. C., *Statistical Techniques in Simulation*, New York: Marcel Dekker, Inc., 1974 (part 1) and 1975 (part 2).

15 LAVENBURG, S. S., AND D. R. SLUTZ, "Introduction to Regenerative Simulation," *IBM J. Res. & Dev.*, XIX, no. 5 (1975), 458–462.

16 LAW, AVERILL M., "Confidence Intervals in Discrete Event Simulation: A Comparison of Replication and Batch Means," Technical Report TR 76–13 (1976), Office of Naval Research, Arlington, Va. (To be published in *Naval Res. Logistics Q.*)

17 NAYLOR, THOMAS H., KENNETH WERTZ, AND THOMAS H. WONNACOTT, "Spectral Analysis of Data Generated by Simulation Results With Econometric Models," *Econometrica*, XXXVII, no. 2 (1969), 333–352.

INDEX